**Site Guides:**

# Costa Rica and Panama

## The Best Birding Locations

### By Dennis W. Rogers

Back cover drawing by Steven Heinl
Other artwork by Anayanci Aguilar B.

For Elena Zamora,
*Tú eres la comemaíz de mi jardín.*

While every effort has been made
to verify the accuracy of the information in this
book, neither the author nor the publisher accept
responsibility for any loss, inconvenience,
or injury that might arise from the
use or misuse of said information.

Cover photo © M. & P. Fogden

©1996 Dennis W. Rogers
ISBN 0-9637765-6-8
Cinclus S.A.
San José, Costa Rica
From overseas:
Cinclus S.A.
SJO 1410
Box 025216
Miami, FL 33102-5216

Printed in the U.S.A.

# Table of Contents

## Introduction
Acknowledgements ............................................. 2
How to Use This Book ....................................... 2
Panama or Costa Rica? ...................................... 3

## Costa Rica
When to Come .................................................. 6
International Transport ..................................... 6
The Airport ....................................................... 7
Domestic Flights ............................................... 8
Rental Cars ....................................................... 8
Buses ................................................................. 9
Driving Standards ........................................... 10
Accommodations ............................................ 11
Food ................................................................ 11
Money & Prices .............................................. 12
Safety .............................................................. 13
The Ticos ........................................................ 14
Language ........................................................ 15
Health ............................................................. 15
The National Parks ......................................... 16
Resources ....................................................... 17
The Cities ....................................................... 18

## The Pacific Lowlands
Carara Biological Reserve .............................. 26
    Birds of Carara ......................................... 28
Rio Tárcoles Estuary ...................................... 29

Birds of the Rio Tárcoles ..................................30
Tivives Mangroves ...............................................31
Manuel Antonio National Park ..........................31
   Birds of Manuel Antonio ................................32
Corcovado National Park ....................................33
   Birds of Corcovado .........................................34
Golfito National Wildlife Refuge ........................37
   Birds of Golfito ...............................................38

## Guanacaste

Santa Rosa National Park ....................................43
   Birds of Santa Rosa .........................................44
Palo Verde National Park ....................................46
   Birds of Palo Verde .........................................47
Barra Honda National Park .................................50
   Birds of Barra Honda ......................................50
Beaches ................................................................51

## The Mountains

Braulio Carrillo National Park ............................54
   Birds of Quebrada Gonzalez ...........................55
Tapantí National Park ..........................................57
   Birds of Tapantí ..............................................58
Volcán Irazú National Park .................................59
   Birds of Volcán Irazú ......................................60
Guayabo National Monument .............................60
   Birds of Guayabo ............................................61
Volcán Poás National Park ..................................63
   Birds of Volcán Poás ......................................64
Virgen del Socorro ..............................................64
   Birds of Virgen del Socorro ............................65
Cerro de la Muerte ..............................................66
   Birds of Cerro de la Muerte ............................68

San Vito ..................................................................69
Birds of the San Vito Area ....................................71
Monteverde ..............................................................72
Birds of Monteverde ...........................................73

# The Atlantic Lowlands

Finca La Selva ........................................................78
Birds of La Selva ..................................................79
Cahuita and Puerto Viejo .........................................81
Birds of Cahuita and Puerto Viejo .......................83
Tortuguero National Park ........................................85
Caño Negro Wildlife Reserve ..................................85
Birds of Caño Negro ............................................85

# Panama

When to Come ........................................................90
International Transport ...........................................90
Domestic Flights .....................................................91
Rental Cars ..............................................................91
Buses .......................................................................91
Driving Standards ...................................................91
Accommodations ....................................................92
Food ........................................................................92
Money .....................................................................92
Safety ......................................................................93
The Panamanians ....................................................93
Language .................................................................93
Health ......................................................................93
Resources ................................................................94
Panama City ............................................................94

# Central Panama

Panama City .................................................. 100
Pacific Side
  Chiva Chiva Road ..................................... 100
  Summit Gardens ....................................... 100
  Plantation Road ........................................ 100
  Gamboa ................................................... 100
  Madden Forest ......................................... 100
    Birds of Pacific Side ............................. 101
  Pipeline Road ........................................... 103
    Birds of Pipeline Road ......................... 105
Atlantic Side
  Achiote Road ............................................ 108
  S-9 Road .................................................. 109
  Tiger Trail ................................................ 109
  Fort San Lorenzo ..................................... 109
    Birds of Atlantic Side ........................... 110
  Tocuman Marsh ....................................... 113
    Birds of Tocuman ................................. 113
  Cerro Azul & Cerro Jefe ........................... 114
    Birds of Cerros Azul/Jefe ..................... 115
  Cana ........................................................ 116
  El Real ..................................................... 116

# Western Panama

Cerro Campana ........................................... 118
  Birds of Cerro Campana ........................... 119
El Copé ...................................................... 119
  Birds of El Copé ...................................... 120
Aguadulce .................................................. 122
  Birds of Aguadulce .................................. 123
Santa Fé ..................................................... 123
  Birds of Santa Fé .................................... 124

# Chiriquí

Boquete ................................................................. 127
Volcán .................................................................. 127
Cerro Punta ......................................................... 128
La Amistad International Park ............................ 129
Santa Clara ......................................................... 129
   Birds of the Western Highlands ..................... 129
Fortuna ................................................................ 131
   Birds of Fortuna ............................................... 132
Chorcha Abajo .................................................... 134
Cerro Colorado ................................................... 134
   Birds of Cerro Colorado .................................. 135
Las Lapas Marsh ................................................ 136

# Bocas del Toro

Oleoducto Road and Chiriqui Grande ................ 137
   Birds of the Oleoducto Road ........................... 137

# Checklist

Checklist of the Birds of Costa Rica & Panama .. 141

# Index

Index of Locations .............................................. 171
Index of Species Mentioned ............................... 173

# Introduction

# Introduction

Costa Rica has justly gained a reputation as a birding destination, and many birders have made their first Neotropical trip to sample the varied habitats there. The small size of the country, large bird list, good field guide, and fast-developing infrastructure combine to make birding relatively easy. The friendly people and general safety of the country means birding is as comfortable as anywhere in Latin America.

Panama is emerging from a period of political instability that made headlines and discouraged foreign visitors. Now is an excellent time to visit this country, with its variety of species, good resources for the birder, and affordable prices. The potential loss of birding habitat in the former Canal Zone means that it may be now or never to bird that area.

## Acknowledgments

This work could not be achieved without the help of many different people. For the Costa Rica section, material by Michael Fogden, Steve Heinl, Jim Lewis, Graham Speight, Gary Stiles, and others was consulted in producing this book, especially the species lists. My thanks to them and all the others who have helped make Costa Rica a world-class birding destination.

I would like to express my appreciation to Dodge Engleman for large amounts of information integrated into the Panama section. Thanks also to George Angher, Lorna Engleman, Craig Faanes, Loyda Sánchez, and Norita Scott.

## How to use this book

Each site in this book merits at least a morning's birding (usually much more), so each stands independent. The book is not organized in any particular fashion (i.e. the "loop routes" unfortunately popularized by the late Jim Lane) to help you plan your trip. Get a map and study the bird lists to decide where you want to place your energy.

The lists are not intended to be comprehensive; for some areas such as La Selva or Pipeline Road that would involve nearly 400 species. They are meant as a study guide for the first-time visitor, to be looked over the night before in preparation for the likely birds of the area. Many of the commonest birds are not mentioned, especially if they are large and easy to identify. Do your homework; obviously the less time you spend with your face in the book, the more birds you'll see.

Some specialties are mentioned, even if they're scarce. These are species that you don't want to miss if you get a chance at them, and are marked with an r after the name. Don't use the lists as a reference to what might occur in the area. See the distribution section of Stiles and Skutch or Ridgely. It's up to you to sort out which species are present if you come during the Boreal summer, and which are not.

I would prefer that the names on the list were in taxonomic order from top to bottom, but despite the large amount of money I paid for this desktop publishing software, it won't allow it! The names are in order, you just have to go from side to side as if you were reading a normal page of type.

A bird-finding guide is no place for innovations in taxonomy or nomenclature, so nothing radical is found here. Specific names closely follow Ridgely and the AOU Checklist, which puts me in conflict with Stiles and Skutch in a few cases: *Turdus* are all called thrush rather than "robin," and "Mistletoe" Tyrannulet is called Paltry Tyrannulet as in most other sources. Names used by the AOU are generally mentioned in the "Notes" section of each species account in Stiles and Skutch, where they differ. *Contra* the AOU, both *Myioborus* "redstarts" are called whitestart by me, and I prefer to lump the forms of Variable Mountain-gem.

## Panama or Costa Rica?

The two countries in this book are actually quite similar from the perspective of the visiting birder. Discounting the relatively inaccessible Darién, the list for each country is in the neighborhood of 850 species. The tourist boom in Costa Rica means that there are more facilities aimed at natural history-oriented visitors, but the roads and other infrastructure in Panama are generally better. Food, hotels, and rental cars are better value in Panama.

Costa Rica has traditionally been more stable and safe than most or all other countries in Latin America, but the situation is worsening. Panama was victimized by the antics of Manny Noriega and his chums for far too long, and is only now starting to recover standing as a civilized country. You still are more likely to be the victim of a violent crime in Panama City (or especially Colón) than in San José, but sadly the latter city now has its daylight muggings too. In the countryside there is little difference in terms of safety. Costa Ricans on the whole are friendlier than Panamanians, but more Panamanians speak English.

The biggest difference from the birder's standpoint is that Panama City is in the lowlands and San José is in the highlands. To get to the good highland birding in Panama, you must drive or fly to Chiriquí, and proceed from there. Likewise, good lowland birding in Costa Rica is at least 1½ hours from the airport. A possible solution would be to combine both countries on one trip. It is virtually impossible to take a rental car across the border, but American and Continental Airlines serve both countries, so you should be able to work out an open-jaw itinerary. The one-way plane fare between Panama and San José is about $125. Consult your travel agent.

# Costa Rica

## When to Come

The best weather for birding is during the dry season, from December to April. This coincides with the vacation season for Costa Ricans; it is especially difficult to travel during the Christmas and Easter holidays. Virtually the entire country heads for the beach during the "Semana Santa" holiday the week before Easter. The rest of the year is rainy, though typically most precipitation comes as a cloudburst in the afternoon, and doesn't affect birding too much. The highlands can be foggy or rainy any time of day or year. Be prepared for mud.

Guanacaste is more pleasant in the "Green" season, as the tourist industry has euphemistically taken to calling it. The trees are leafed out and wind isn't such a problem. Roads there can be bad, but that isn't much of a factor for the areas covered.

On the Caribbean slope it can rain at any time. During the dry season, northern weather systems sometimes produce rainy spells of several days. If you find yourself in one of these "temporales," it may be worth changing your itinerary to visit another part of the country until the weather is better.

Temperatures in the highlands are generally moderate, though it can be chilly and windy at the highest elevations. The lowlands in the dry season are hot and muggy, except in Guanacaste where it is hot and dry.

The dry season corresponds to winter for numerous migratory species. These can make for bigger flocks and more variety, and are of special interest to the birder from western North America, since the majority of migrants are from the eastern side of the continent.

## International Transport

The main international airport near Alajuela is well-served from the U.S. and Canada, while continuing development in Guanacaste means the newly-expanded airport at Liberia will receive more flights. LACSA is the principal Costa Rican airline, serving New York, Miami, Orlando, New Orleans, Los Angeles, and San Francisco. It has a reasonable reputation and is a good option if you live near one of those cities. Recently Aero Costa Rica started service to Miami, Orlando, and Atlanta.

U.S. airlines serving San José include American from Miami and Dallas/Ft. Worth; Continental from Houston; and United from Los Angeles via San Salvador and Washington via Mexico City. There are of course good connections to most cities in the U.S. from these hubs. Canadian also serves San José from Vancouver seasonally, and there are charters from eastern Canada.

Travelers from Europe can fly with Iberia or KLM but most connections are via Miami.

Other Latin American airlines also may offer good fares, though only Copa is considered reliable. Connections to South America can be made via Panama, Venezuela, or Colombia. There are also direct flights to Ecuador, Peru, and Chile.

It is also possible to reach Costa Rica by highway, but it is not recommended. The expense and the hassle of crossing so many borders make this route less than worthwhile. In addition to the obvious travel expenses, you will have to pay so many fees and bribes (El Salvador and Honduras are notorious in this regard) that it can be cheaper to fly. Allow at least six days from Brownsville if you decide to drive, and don't tell the Mexicans you are going to Central America.

## The Airport

The Juan Santamaría International Airport gives most travelers their first impression of Costa Rica, and often it is not a good one. Departing and arriving passengers mingle freely in the concourse on the airside of customs. Follow the signs to customs and immigration, to the right from most gates. Some charters arrive at a remote terminal. With the scheduled carriers, you will then go down the stairs to the long lines at immigration. Nationals of most developed countries need only a valid passport to stay up to 90 days.

Once you've made it through immigration, your bags should be arriving on one of the two carrousels. If you have two or fewer bags, you can pass through the "green" line, where a "stop light" is rigged to randomly chose people for inspection. Most Gringos go through with a minimum of hassle.

Outside there will be a mob of people greeting their friends and relatives, along with the usual taxi drivers and hotel touts. There is also an official tourist information office. The airport bank is inside the departure terminal, but it keeps regular government banking hours and is not too useful. The rental car kiosks are in front of the terminal.

You are still about 20 minutes (depending on traffic) from downtown San José. The taxi service from the airport is a monopoly, so prices aren't too good. A regular bus stops by the terminal, useful if you don't have much luggage.

Departure is usually a little easier. As you enter the terminal, the U.S. airlines are to the left and LACSA to the right, with the others in the middle. Lines can be long so allow plenty of time. After you're checked in, there will be another line for immigration. Along the way to the plane your boarding pass will be checked at least three times. Note that there are actually two entrances to immigration and security, one on each end of the terminal. LACSA doesn't have too many flights in the morning, so if you are leaving early, it's usually worth checking that things aren't less congested at that end.

Be sure to allow enough money for your airport departure tax, which is NOT included in your ticket the way it is in the States. For tourists the tax is presently $17 or its equivalent in Colones, payable at the window in the corner or to the airline people. It is generally better to pay in Colones as the rate tends to lag a little behind devaluations.

Recently a number of hotels have cropped up near the airport, and there are others in easy-to-reach parts of Alajuela or San José. These can be convenient for an early departure.

## Domestic Flights

There is an ample schedule of domestic flights but they are not all that useful, as the birder will need transport at the end of the flight, and rental cars are not widely available outside San José/Alajuela. SANSA is government-owned and has a reputation for low prices and unreliability. Travelair is more reliable at higher prices. Other than a possible visit to Tortuguero with Travelair, the birder should stick to the highways. Travelair flights depart from the domestic airport in Pavas, on the west side of San José.

## Rental Cars

The last few years have seen an explosion of small rental car companies in Costa Rica. Competition and favorable tax treatment have brought prices down to the $250 per week level for a small car with unlimited milage. Four-wheel-drive can be had for as little as $300 per week. Some of the small companies are competitive, but I prefer the international ones like Avis, National, Dollar, or Budget, because if they pull something on you then at least you can complain at home. Be sure to shop around and reserve from your home country for the best prices. See if you can waive the overpriced, high-deductible standard coverage; they will often tell you it's mandatory. Take your time and read the small print. But beware that you may be liable for any problems, even mechanical breakdowns unlikely to be covered by a card's insurance.

If you do have an accident, don't move the car until the police say you may. Be sure you have liability coverage. In the event of an accident where someone is hurt, you could be prevented from leaving the country or even jailed.

When you get your car be sure to check it for scratches and dings, so you don't get charged for damage that was already present. If you get a flat by all means get it fixed, as the car companies may charge up to 10 times the gas-station rate. If you want to go to Monteverde, make sure your contract allows you to use that road; some don't. Others prohibit driving on any unpaved road, so a 4WD may be worth the extra money.

The worst problem with rental cars is that most have special licence plates that are an invitation to trouble from thieves and corrupt traffic police. Break-ins are a special problem around Carara and Jacó, in Braulio Carrillo, and of course in San José. Don't leave your car on the street in San José, put it in a parking lot (about $1/hour). Don't leave anything of value in view from outside, anywhere.

Most small rental cars are Japanese models with trunks which could be used to store valuables out of sight, except most have an internal release. A thief need only break a window to have ready access to the trunk. I suggest you bring a pair of needlenose pliers.

Gasoline prices in Costa Rica are fixed by the government at about $1.65 per gallon; Colon prices change regularly with devaluation.

## Busses

Busses go almost anywhere you might want to go in Costa Rica, and they are cheap. The longest run in the country, eight hours of pure suffering from San José to Puerto Jiménez, is only about $6. If you have the time and have no one to share costs, this is a good alternative. For specific information about schedules, visit the Instituto Costarricence de Turismo information office under the Plaza de la Cultura in San José.

Nonetheless, access to the good birding spots is limited. Some areas have bus service, but it is often difficult to arrive early unless you are willing to camp in the habitat. Others locations would require long walks to get to the birding area. Spots in this book accessible by bus include: Monteverde, Carara, Manuel Antonio, Golfito, San Vito/Las Cruces, Cerro de la Muerte, Guayabo, La Selva, Virgen del Socorro, and Cahuita/Puerto Viejo. Some other areas, such as Volcán Poás, are close enough to settled areas that taxi service is reasonably priced. For more information on this sort of travel, see the Lonely Planet guidebook.

## Driving Standards

Costa Ricans tend to lose their agreeable personalities behind the wheel of a car, and that combines with the difficult roads to make driving an adventure. Highways outside the Central Valley tend to be two lanes of asphalt with no lines or lighting. The winding roads from the valley down to the coasts, and over the Cerro de la Muerte, are especially difficult. Night driving is not recommended due to potholes, animals in the road, and vehicles without lights. If you need to drive at night to reach a birding area early in the morning, drive slowly and carefully. Some idiosyncracies of San José are treated under that section.

Another problem is the traffic police, well-known for their corruption. They may seize on some real or imagined offense to try to extract a bribe from you, telling you that you have to go to an inconvenient place to pay the fine. The intersection speed zones along the freeway from the airport to San Ramón are notorious for their radar traps. The proper thing to do is just take the ticket and give it to your rental car company. They will pay it and add it to your bill (possibly with a surcharge). If you have been unfairly ticketed, it's possible to complain. Unless you have Spanish-speaking witnesses, nothing will happen. Even if you have witnesses, probably nothing will happen. The officer's badge number and signature are on the ticket. Any policeman in Costa Rica should show i.d. when asked. Of course if he's already doing something illegal, he is unlikely to be swayed by that rule.

If your Spanish is up to it you can pay off the traffic cops, waiting for the suggestion that you pay the fine now. Often this will be too subtle for you to realize it, though with tourists they will be more blunt. The going rate to fix a speeding ticket is about 1,000 Colones, so don't let them bully you into more than twice that. Most of the fines are in the neighborhood of 5,000 Colones (plus an additional 30% for the Abandoned Children's Fund), so it is an economical proposition to pay them off. Don't worry about corrupting the natives: due to low pay (about $250/month), poor supervision, and a general lack of professionalism, they are already corrupt.

Traffic police must stand within view of their cars (various small pick-ups painted blue and white and Toyota sedans painted light blue) or motorcycles. Often they are wearing orange vests as well.

## Accommodations

Accommodation in Costa Rica runs the full gamut, from first-class hotels to concrete-block hovels that suggest prison cells. San José has the best selection, but in recent years lots of nice places catering to tourists have sprung up in the countryside.

Value for money has suffered during the current tourism boom, as hotel owners have jacked up prices with no corresponding increase in quality. Generally in birding areas $30-50/night for two people will give you an ample selection. Beware some really cheap places that are also brothels (prostitution is legal in Costa Rica). Any place called a "motel," and/or where you cannot see the parking from the road, is usually a short-time place for consummating illicit affairs.

During the peak season it's perhaps a good idea to make reservations, but it's not really necessary most of the year. As of 1996 it was clear that the tourist industry had overbuilt, and accommodation is now somewhat in surplus. If you insist on a certain standard of accommodations, come in the off season or book well ahead. See a good tourist guidebook and/or make reservations with INFOTUR, across the street from the Teatro Nacional in San José (506-223-4481, fax 223-4476). Other travel agencies can also book hotels, but this company seems to have the best variety.

## Food

Cuisine is not one of Costa Rica's strong points, though "typical" food is usually good value. Rice and beans are the staple, supplemented with tough but tasty meat, overcooked vegetables, and a wide variety of fruit. Outward appearance is not a reliable measure of a restaurant. Some shabby *sodas* have excellent fixed plates (*casados*), while other fancy-looking places have poor food. One can eat modestly but well on $10/day even in San José, though prices tend to be higher at the beaches.

Visitors trying to maintain a low-fat diet will not get much help from the locals, who like their fare oily. The use of animal fat for frying has mostly died out, with hydrogenated vegetable oil taking its place. It has taken the same word, however, so if your dictionary translates *manteca* as "lard" don't panic. The distinctive Caribbean cooking unfortunately relies heavily on cholesterol-laden coconut oil.

## Money & Prices

The Costa Rican currency is the Colon, as of September 1996 at about 213 to the U.S. Dollar, with creeping devaluation. Through most of 1996, the rate was rising at about half a Colon per week (1% per month) as part of a policy intended to fend off shock devaluations.

Wait to change your Dollars until you arrive, as rates in Miami are poor. Currencies other than U.S. Dollars are a problem, though Canadian Dollars can be exchanged (most easily at Scotiabank downtown). Due to recent liberalization of exchange rules, life is much easier for the tourist. Lines in the banks can still be horrific. At the government banks (Banco de Costa Rica, Banco Nacional de Costa Rica, and Banco Crédito Agrícola de Cartago), a couple of hours for a routine transaction is not uncommon. You will generally have to wait in one line for paperwork, and then another to actually change your traveler's checks or cash. Show your money to the guard who will point you towards the right line. A better choice is one of the private banks, where the same transaction will take a few minutes at most. Unfortunately, these are concentrated in the San José area, though several are expanding rapidly. Most of the better hotels will change money for guests. Leftover Colones can be easily changed back to Dollars.

Don't change on the street, as exchange reform has snuffed out the black market, and the money changers that remain are scam artists. If you arrive late at night, on a Sunday, or on one of the many government holidays, the bank at the airport will be closed. Somebody will probably approach you and offer to change at a reasonable rate. In this environment and with care, it is acceptable to make the transaction. To avoid slight-of-hand, watch carefully and don't hand over your dollars until the Colones are counted out.

Traveler's checks are the best way to bring money if you can get them free; otherwise cash advances on a VISA card are relatively easy to obtain (2% commission) at branches of the Banco Nacional, Banco Crédito Agrícola de Cartago, or several private banks.

Credit cards are gaining acceptance in Costa Rica, with VISA the best choice and MasterCard also widely used. Your card is less likely to be accepted in the countryside. To avoid fraud, it is best not to let the card out of your sight. Of course, check your bill carefully when you get home.

Costa Rica is not a particularly cheap country to visit. Inflation of 18-20% per year in Colon prices has been mostly made up by the devaluations, but Dollar prices are still increasing. Hotel prices especially have been inflating rapidly despite lagging occupancy; overpricing of tours, food, and rental cars plagues the tourist industry.

Once your plane ticket and rental car are paid for, two people traveling together should be able to get by in normal middle-class fashion on about $40/day per person. Even traveling by bus in bare bones fashion, getting by on much less than $20 per day is difficult. It certainly can be done if you come in the dry season and camp.

## Safety

Aside from some specific areas that will be mentioned in the text, your personal safety is not really a concern anywhere in Costa Rica. (Property is another matter.) People in the countryside are almost never a problem.

In the city, beware of most of the downtown and especially the "Coca-Cola." Generally the quadrant with even-numbered *Calles* and odd-numbered *Avenidas* is considered to be the rough side of town. Unfortunately, many of the long-distance bus stations are in that area.

Remember, even if you don't consider yourself rich, many Costa Ricans can and do live for a year on what you'll spend on this trip. This makes you a target for any pick-pocket or snatch-and-run artist who sees you. Don't help them out by wearing expensive jewelry or watches, and if you must carry that camcorder around town keep a firm grip on it. Keep your wallet in a front pocket.

Even with precautions, there is still a good chance of losing something, as petty theft is very common. Not only are there more professional pickpockets and bag-snatchers than in the U.S. (for example), but there are many opportunistic thieves who won't pass up the chance to augment their possessions at your cost. Nonetheless, the chance of a serious encounter is as low in Costa Rica as anywhere I know.

Other than *Homo sapiens*, insects are probably the most serious concern for the birder. Get used to looking for wasps, bees, and especially ants before you put your hands or feet anywhere. Rubber boots are a good idea in the forest. The sting of the large *bala* ants found in the Caribbean lowlands is reputed to be especially painful. Africanized honeybees are another problem, so keep an eye out for swarms of bees, especially in Guanacaste. If you come upon a nest or swarm, sneak away quietly.

Chiggers can make your life miserable. They are most prevalent in lowland areas with livestock, but are found in the forest as well. Treating socks with DEET or sulphur powder has been suggested, but nothing seems to work really well.

Snakes are much overrated as a threat in the tropical forest. Most poisonous snakes are nocturnal, and unlikely to be encountered while birding. If you stay on the trails and watch where you're walking, there is no need to worry. In the event that you are bitten, try to get a look at the snake involved for identification, and then go calmly to the nearest town for treatment.

Monkeys can also be a minor hazard in the forest. Your interest in them will most definitely not be reciprocated, and given the chance, they will throw sticks or worse at you.

## The Ticos

Costa Ricans are very friendly, and can usually be counted on to do their best to help you out. They are genuinely fond of foreigners.

Decorum is especially important with Latins in general and Ticos in particular, so no matter how bad the service or how ridiculous the directions you're given, it is essential to maintain your composure. Under no circumstances should you show indignation or anger.

Ticos are not too good at giving directions, though at least you probably won't be sent off in the wrong direction on purpose. Most Costa Ricans live their whole lives in a small area, and cannot relate to the type of directions someone unfamiliar with the local landmarks will need. San José has a perfectly logical grid system of streets and avenues, but virtually no one uses it, instead saying "X meters in such-and-such direction, then X meters in another" from some well-known building, park or other landmark. The occasional use of landmarks that no longer exist (like the old U.S. embassy downtown) makes it even worse for the visitor. Maps do little good in this situation. Usually *cien metros* (100 meters) means a city block regardless of the actual distance, but sometimes you will run into someone who is talking about the physical distance. In recognition of the problems this situation causes the tourist, most of downtown San José now has signs on street corners. With luck you can then get an address based on the street numbers.

When away from San José, try to ask directions of someone who looks like he/she drives. Taxi drivers are good sources. Often you can simply ask *a* (to) your destination and try to get the person to point. If he points one way and says another (*izqierda/derecha/directo* for left/right/straight) go with the finger.

## Language

Spanish is the only language of the vast majority of the population, so outside of the better tourist facilities don't expect anyone to speak English. The minor exception is the Caribbean coast, where some of the locals speak a Creole English. Menus are often bilingual.

Before you leave home, work on some phrases you can use when you get lost—you doubtless will despite my best efforts.

## Health

Costa Rica is one of the healthiest countries in Latin America, due to its relatively advanced standard of living and high social spending. Emergency care is free for everyone in Costa Rica, foreigners included. Your principal health risk will be a traffic accident.

Traveler's diarrhea is almost inevitable. Bring Pepto-Bismol for discomfort and Immodium if you have to leave the security of your hotel bathroom. You might avoid this malady by not drinking any water (don't forget ice), especially in the lowlands, and by eating only hot food that you can see being prepared. On any trip of more than a few days you will certainly be exposed to the foreign strains of *E. coli* that cause most problems, so you might as well resign yourself to it. Severe parasites like amoebas and *Giardia* are much rarer, but present.

The warm and moist climate of the tropics is ideal for the reproduction of the bacteria that cause food poisoning, and many restaurants do little to cut the risk. In most cases your best strategy is to stick to hot food produced in reasonably clean surroundings. Often foods like fried chicken are kept warm with a light bulb; don't eat them.

Beer and soft drinks are fine, as are coffee or tea once the water has been boiled. Commercial milk products are no problem.

Malaria is on the rise in the Atlantic areas of Costa Rica, though still relatively rare. That isn't to say that you should or should not take prophylactic medication, as it can have side effects. Consult your physician. A more serious problem in 1994-95 was a Dengue fever outbreak in the Puntarenas area, with a few cases elsewhere on the Pacific coast and in the Central Valley. The only measure against this disease is to avoid being bitten by nocturnal mosquitoes.

## The National Parks

Many of the spots in this book are protected and kept public within Costa Rica's fine system of national parks. Technically most of the parks don't open to the public before 8 a.m., but in many cases one can sneak in or just amble by and pay your entrance later. The administrators generally realize that such a late hour is not acceptable to many tourists, but the bureaucratic, centralized system is slow to change.

As of August 1994 the entrance fee was 200 Colones, up from 25 Colones five years before. The government then decided to try to pay for the operation of the parks with admissions alone, and the rate for foreigners went to $15, with a complicated system of advance discounts. A year of protest from the industry and problems with corruption among the park guards (unsurprising when each foreigner is paying a day's wages for the ticket taker) ultimately resulted in a reduction to $6 for all parks. Hopefully this will be the end of the story.

Some of the extra money has at least gone to improve the generally poor information and maps available. If you pay the full rate you will perhaps now be given a map, at some of the major parks at least.

There is also a program for international volunteers at the various national parks. Participants work at maintaining trails, patrolling boundaries, etc., and stay at the park stations. They will make you jump through some bureaucratic hoops and require a commitment of 45 days, though it would be impossible to enforce this. Going as a volunteer is a good way to get to know the parks, if you have plenty of time. Usually the only cost is a small fee for food. For more information, drop by the office at Avenida 2 and Calle 19 or write to Oficina de Voluntariado del Servicio de Parques Nacionales, Apt. 11384-1000, San José. The phone there is 222-5085, Spanish only.

## Resources

The main source for birders is Stiles and Skutch's *Birds of Costa Rica*, with Ridgely's *Birds of Panama* second. Stiles and Skutch is a difficult book to use due to its detail; there is a vast quantity of information that must be waded through to reach what you want. It is also a bit heavy to carry in the field, so it is well worth getting the plates bound separately. Ridgely's book is better for beginner's identification problems since he discusses similar species under each account, but descriptions of voices are better in the Costa Rican book. The quality of the artwork is similar in both books. The second edition of Ridgely (1989) has accounts and illustrations for all the species found in Costa Rica but not in Panama, but these are in a separate section, and there is no Costa Rican locality information for birds also found in Panama. Behavior and habitat information are much better in Stiles and Skutch. In other words, bring both books. A Mexico book is useful as well, especially in Guanacaste, but is not really necessary. The new Howell & Webb book is highly recommended.

A few other publications are available, including two checklists. *Birds of Costa Rica, Locational Checklist* has abundance info for 17 locations around the country. The abundance codes are a bit optimistic for the inexperienced birder, but it provides a good idea of where to look for particular species. *An Annotated Checklist of the Birds of Monteverde and Peñas Blancas*, by Michael Fogden, gives detailed abundance information for the various zones of that well-known location.

You will need a map, of which several are available. None is particularly accurate when it comes to detail. The best, and probably most widely available in North America, is the International Travel Map Productions *Traveler's Map of Costa Rica*. Good, but expensive, is *The National Parks and other Protected Areas of Costa Rica*, produced by the Fundación Neotrópica of San José.

For general tourist information, the Lonely Planet guide *Costa Rica: a travel survival kit* and *The New Key to Costa Rica* (Ulysses Press) are generally the best choices. *Costa Rica: A Natural Destination* has some information for birders. The web-based *Green Arrow Guide to Central America* has information on ecotourism and birding destinations, at http:\\www.greenarrow.com.

The ICT tourist information office, under the Plaza de la Cultura in downtown San José, is very helpful with specific questions.

## The Cities

Getting around San José will be the first, and perhaps most memorable, adventure of your trip. Infrastructure construction just isn't keeping up with demand. Costa Rica's capital doesn't yet have the smog of Mexico City, the bad drivers of Guatemala, or the gridlock of San Salvador, but with the present rate of economic growth and vehicle importation, it's gaining fast. Be prepared to get lost and keep your humor as much as possible.

Most signs are of the international type with the possible (and important) exception of "No hay paso" which means you are entering a one-way street the wrong way. Usually oncoming traffic will emphasize this for you.

Stop signs are treated liberally by the natives. Where there is a stop sign and a traffic light at the same intersection, the light has priority. Usually there is only one light for each direction, so when a bulb is burned out (often) it may appear that the light is not functioning. Watch the other traffic for a cue; usually the signal will change and light up again. Especially on the highway eastbound towards Cartago, there are places with two lights. One controls left turns and the other is for through traffic.

The traffic circles on some San José byways can be difficult to get the hang of if you don't have them at home, but the general rule is to enter in the left lane if you are going through or left, and in the right lane for the short right turn. Then yield to anyone in the circle before taking your place in the controlled chaos.

Generally the birder will want to stay at a hotel on the edge of town. There are a number of luxury hotels now on the west side of San José. The Hotel Irazú is often part of discount air and hotel packages. It is also a landmark in the following discussion of how to cross the city.

If you are leaving early in the morning it will not matter much which route you use, though navigating without daylight to read the (few) street signs can be a problem. Generally, when traffic is light the downtown route is better due to its simplicity. When daytime traffic sets in the northern route through the industrial suburb of La Uruca is best.

**City Route:**

Technically, the Pan-American highway passes right through the center of San José, and some signs direct you that way. If coming from the airport and heading towards Cartago, you basically want to continue on the freeway past the Irazú and Corobicí hotels, and then take a left at a stoplight just past an Exxon station and opposite a large park. The turn is well-signed, and much of the traffic will go this way. This will put you on the Paseo Colon, which is actually Avenida Central (0). Continue and jog right with most traffic onto Avenida 2 as you enter the downtown, then continue straight about 5 km to the suburb of San Pedro, where you will see a traffic circle with some upside-down fountains in the middle. Go around the traffic circle and straight to the suburb of Curridabat. After another 4 km you will go over the Cartago freeway. Be sure to get in the middle lane (marked "Tres Ríos/Cartago") or you're in trouble. Merge as soon as possible, at the beginning of what appears to be a long merge lane but is not. Shortly after that, you will need to pay a toll (outbound only), 60 Colones in 1996. The two left lanes here are automatic machines, for which you need 10 or 20 Colon coins. The other one or two booths will have a human dispensing change and tokens, though you may still have to feed the machine.

Westbound on the same highway, don't take the freeway to its end. Take the exit for San Pedro and San José, and return to the fountains mentioned above. As you continue towards the city you will be eventually confronted with a "No Hay Paso" sign and be forced right. Take the first available left, and continue towards the downtown on Avenida 1. On through downtown, either jog left on Calle 20 or right on Calle 22 and continue westward until you hit the airport freeway.

**Southern route:**

This beltway is actually the best way to get from the west side to the highway to Cartago, but its many traffic circles are miserably signed. If you're feeling adventurous and want to try it, the general rule is that you go straight at each circle.

**Northern route:**

To avoid downtown traffic, when coming from the airport look for some strange signs which show the convoluted route to the Irazú hotel. After seeing these and a large hospital (CCSS) on the right, pass under a pedestrian overpass and then exit right before the freeway goes up onto a flyover. On the traffic circle below, go nearly all the way around (until you are almost facing back the way you came) and exit, getting in the right lane. After a few blocks you will see the large LACSA administration building on the left, and will want to turn right at a major intersection. This will take you through La Uruca, where most of the heavy equipment and car dealers in Costa Rica are. After about 3 km there is a fork with a Shell station in the middle: take the left. This continues straight (except for one jog to the right), until you reach an intersection with a sign for "Circunvalacion Sur" and bear right. After two traffic circles (go straight), you will need to exit right at a large shopping center, just after Taco Bell, to get to the San Pedro traffic circle. Around the circle and left you will be on the highway to Curridabat, as in the other eastbound directions above.

Westbound this route is far superior. At the San Pedro circle, go right onto the beltway and through the first two circles. At a T intersection you will need to make a left, then continue straight for about 5 km to the fork with the Shell station, where you merge to the right. At the intersection by the LACSA building, you can then go straight to the freeway and airport, left to the Hotel Irazú, or right to Heredia for Volcán Poás, Virgen del Socorro, etc.

**To Braulio Carrillo and Guápiles:**

It would seem that the route just described would also be the way to get on the Guápiles highway for Braulio Carrillo, Finca La Selva, or Limón, but there are no ramps where it crosses the freeway. From the airport, it is probably easiest to work your way into the downtown grid and get on Calle 3 northbound, which will eventually turn into the Guápiles highway.

An alternative route is faster but very difficult to describe. It should probably not be attempted in the dark, at least until you are familiar with it. Follow the directions for the northern route to Cartago. Take the left at the Shell station mentioned above. The road to the left, at the next stoplight, is the one you want, but you cannot turn left at this intersection. To beat this new control, some of the traffic will go to the next corner and turn left into the residential neighborhood. Another left and a right, and you are on the correct road. After going down and back up the hill, you will cross the railroad tracks by Metalco, where much of the corrugated metal roofing in Costa Rica is made. Bear right with the main road, then left at a fork, and then straight through a stoplight. Shortly you will see the buildings of La Nación, Costa Rica's largest newspaper, on the left. After another curve, you will come to a bridge with two bushy pine trees growing by the near end. This is the highway overpass. The ramp for Guápiles is on the right across the bridge. If you get lost ask for *el puente de Llorente*.

From Cartago it's much easier to get on the Guápiles highway. Bear right at the San Pedro circle, and pass through the next one to the

Guadelupe circle, just after the Lubriquick and a large cemetery. Go around the circle and left. At the fifth traffic light, by the La Republica (a newspaper) building, turn right onto the highway.

**Cartago:**

Costa Rica's old capital doesn't present nearly as much of a challenge to the visitor. For Paraíso (Tapantí National Park) and Turrialba, stay on the freeway down off the continental divide at Ochomogo, until two lanes curve off to the left and one goes straight (to Cerro de la Muerte and San Isidro). Go towards Cartago (left). After a large cemetery at the edge of town, go right at the first opportunity. After two blocks take a left, and follow that street all the way through the city to Paraíso.

Westbound it is a different set of streets but the same idea. The route is fairly well-signed for San José.

**Heredia:**

In order to reach Volcán Poás and Virgen del Socorro it's best to pass through the University town of Heredia. From San José via the LACSA building, you will arrive at the southeast edge of town. Once in town, the main drag continues uphill past the National University, and then ends at a T. Take the left, following signs for the Cafe Britt tour. After a few blocks there is another Britt sign, where you want to turn right onto the highway to Barva and eventually Puerto Viejo.

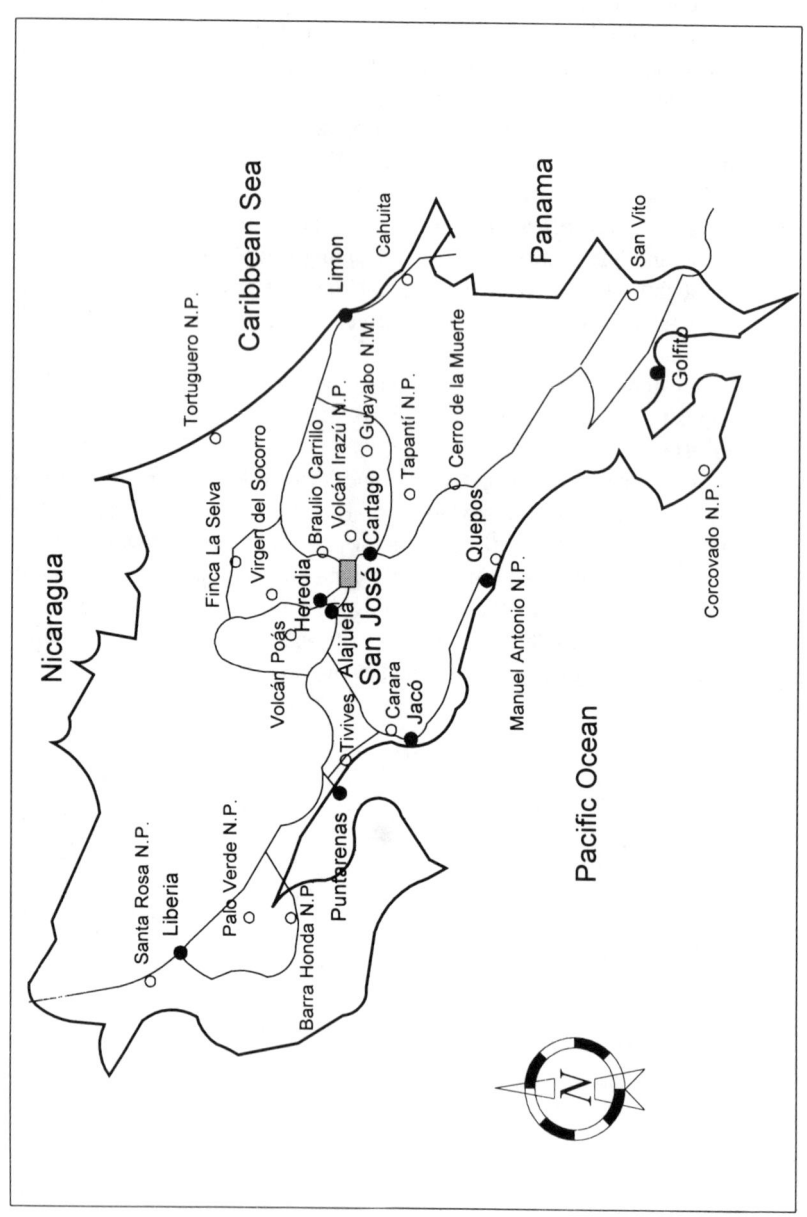

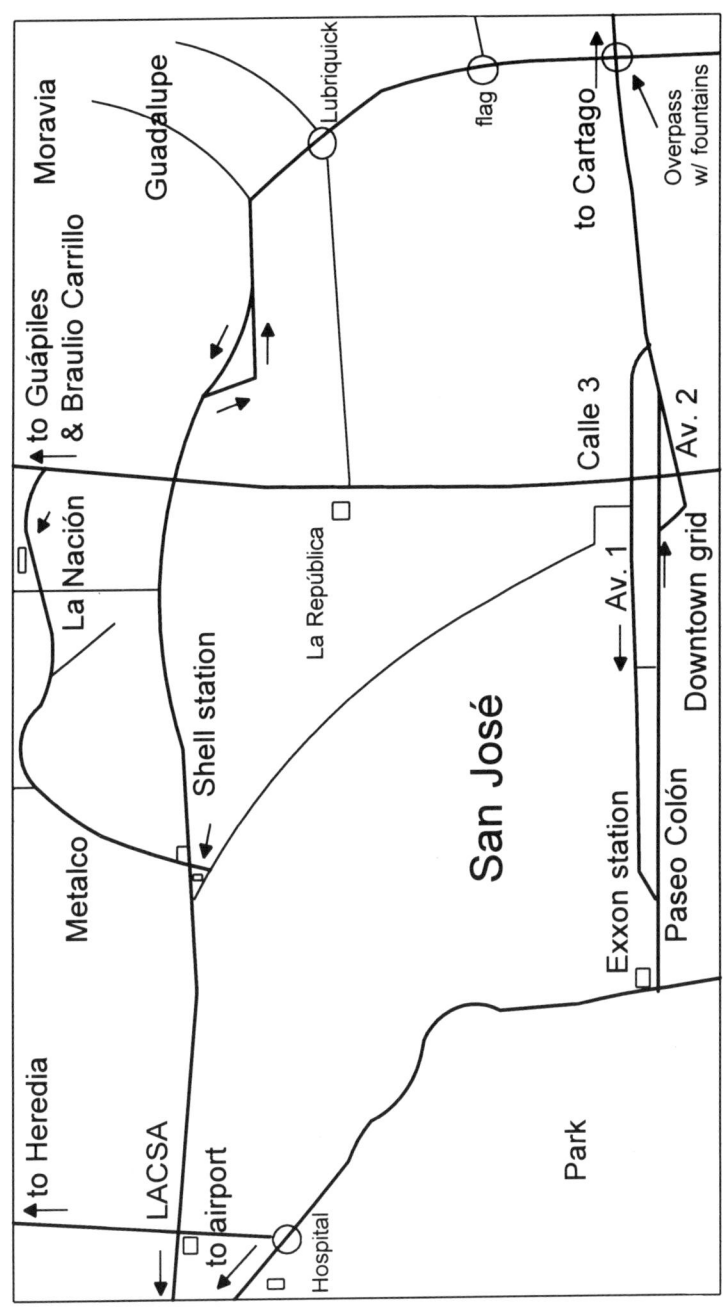

# The Pacific Lowlands

# The Pacific Lowlands

The narrow band of moist lowland forest from Carara southeast into Panama contains some of the best birding in Costa Rica. Unfortunately, the areas of undamaged forest are getting smaller and more difficult to reach. A number of localized endemics make this zone essential for the visiting birder.

This area also has some of Costa Rica's nicest beaches as a diversion. This has both good and bad effects, as the proximity of San José means big crowds at Jacó, Quepos, and Manuel Antonio on weekends. Schedule accordingly. Accommodations are well-placed and plentiful as a result of the demand, though prices can be high in the "summer" season. Easter weekend is out of the question. Good value can be had in the off-season, however, especially in Jacó or Quepos.

## Carara Biological Reserve

This reserve, while not a national park per se, provides the easiest access to good Pacific lowland forest. A reasonably large population of Scarlet Macaws is the principal attraction, but a wide variety of other species can be seen. The trails can fill up with tourists, so arrive early.

The reserve marks the transition between the Tropical Dry Forest and the wetter forests of the southern Pacific coast. A substantial number of species reach the northern or southern limit of their ranges here. The area by the Río Tárcoles is a good place to start the day with a few open country species. The crocodiles that are usually beneath the bridge are a major attraction.

The accessible forest is right on the coastal highway near the beach town of Jacó, about two hours from San José. Take an exit from the freeway about 11 km beyond the airport, and pass through Atenas and Orotina. About 300 meters beyond the Río Tárcoles bridge is a road off to the left into the forest. This is the main birding area. This area has a contingent of professional thieves and has suffered many car break-ins. If there are no tour bus drivers or park rangers around to watch your car, you should leave it at the restaurant on the other side of the bridge and walk (good birding anyway) to the entrance.

There are also some short trails through nice forest at the headquarters, about 3 km along the road toward Jacó.

Many tourists visit this area as a day trip from San José, but you would need a very early start to make this worthwhile for birding. You'll want to be well down the trail before the dudes show up. There is plenty of accommodation in Jacó and a few cabinas and one lodge in Tárcoles/Playa Azul.

The Scarlet Macaws can be easily seen early or late in the day as they fly between the reserve and roosting areas in mangroves to the northwest. During the day you might get a look at a pair feeding in the reserve.

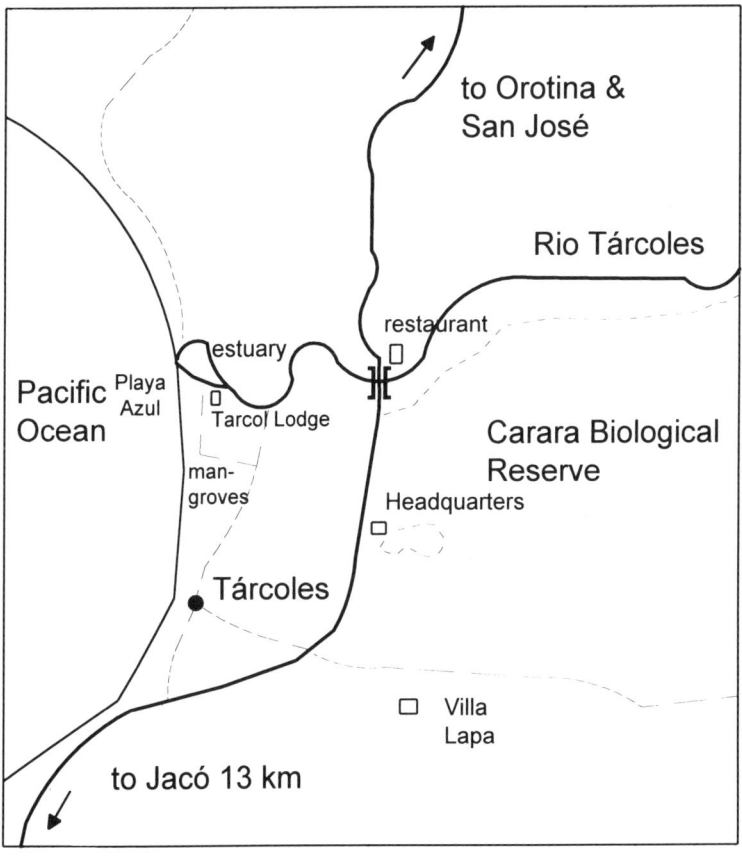

# Birds of Carara Biological Reserve

Great Tinamou
Broad-winged Hawk
Yellow-headed Caracara
White-throated Crake
Inca Dove
White-tipped Dove
Crimson-fronted Parakeet
Orange-chinned Parakeet
Red-lored Parrot
Yellow-naped Parrot
Lesser Nighthawk
Bronzy Hermit
Long-tailed Hermit
Scaly-breasted Hummingbird
Blue-throated Goldentail
Purple-crowned Fairy
Black-headed Trogon
Violaceous Trogon
White-whiskered Puffbird
Fiery-billed Aracari
Lineated Woodpecker
Buff-throated Foliage-gleaner
Tawny-winged Woodcreeper
Barred Woodcreeper
Streak-headed Woodcreeper
Barred Antshrike
Dot-winged Antwren
Black-faced Antthrush
S. Beardless Tyrannulet
Northern Bentbill
Eye-ringed Flatbill

Gray Hawk
Crested Caracara
Crested Guan
Short-billed Pigeon
Blue Ground-Dove
Gray-chested Dove
Scarlet Macaw
White-crowned Parrot
Mealy Parrot
Squirrel Cuckoo
Band-rumped Swift
Band-tailed Barbthroat
Little Hermit
Violet-crowned Woodnymph
Rufous-tailed Hummingbird
Long-billed Starthroat
Baird's Trogon
Slaty-tailed Trogon
Rufous-tailed Jacamar
Golden-naped Woodpecker
Pale-billed Woodpecker
Plain Xenops
Wedge-billed Woodcreeper
Buff-throated Woodcreeper
Great Antshrike
Black-hooded Antshrike
Chestnut-backed Antbird
Spectacled Antpitta
Ochre-bellied Flycatcher
Common Tody-Flycatcher
Yellow-olive Flycatcher

| | |
|---|---|
| Golden-crowned Spadebill | Royal Flycatcher |
| Ruddy-tailed Flycatcher | Black-tailed Flycatcher |
| Yellow-bellied Flycatcher | Bright-rumped Attila |
| Rufous Mourner | Dusky-capped Flycatcher |
| White-winged Becard | Rufous Piha |
| Yellow-billed Cotinga | Three-wattled Bellbird |
| Thrushlike Manakin | Orange-collared Manakin |
| Blue-crowned Manakin | Red-capped Manakin |
| Gray-breasted Martin | Mangrove Swallow |
| Black-bellied Wren | Riverside Wren |
| Rufous-breasted Wren | Plain Wren |
| House Wren | Long-billed Gnatwren |
| Tropical Gnatcatcher | Yellow-throated Vireo |
| Yellow-green Vireo | Lesser Greenlet |
| Tennessee Warbler | Chestnut-sided Warbler |
| Bananaquit | Golden-hooded Tanager |
| Scarlet-thighed Dacnis | Blue Dacnis |
| Green Honeycreeper | Red-legged Honeycreeper |
| Yellow-crowned Euphonia | Thick-billed Euphonia |
| Red-crowned Ant-Tanager | Summer Tanager |
| Buff-throated Saltator | Blue-black Grosbeak |
| Orange-billed Sparrow | Black-striped Sparrow |
| Blue-black Grassquit | Variable Seedeater |
| White-collared Seedeater | Northern Oriole |
| Yellow-billed Cacique | Scarlet-rumped Cacique |

## Río Tárcoles Estuary

While unlikely to produce many species not found in North America, this is the most convenient spot to pad your list with terns, herons, and a few migrant shorebirds. Mid-tide is best, so check the tide tables on the last page of *La Nación*. The tides at Puntarenas should be close.

To reach the best area for viewing, take the first right after the Carara reserve headquarters, clearly marked for Tárcoles. At a T intersection in town, go right back towards the north and continue about 3 km to Playa Azul. Bear right to avoid the beach, and the estuary will shortly come into view. It is also possible to reach the mouth of the river by walking across the soccer field to the beach, and then through some mangroves to the estuary. The soccer field is about a 100 meters back towards town from where the road hits the river.

The sandbar at the mouth of the river usually has a flock of gulls, terns, pelicans, and Black Skimmers. The whole area can be good for shorebirds and herons, depending on the tide. Boat-billed Herons roost along the river upstream and can be seen from a good vantage point such as the one near "Crocodile Jungle Tours" dock.

The patch of mangroves on the road in to Playa Azul might produce herons, Mangrove Black-Hawk, Mangrove Hummingbird, Yellow-billed Cotinga, Mangrove Vireo, or other mangrove specialties.

## Birds of the Río Tárcoles Estuary

| | |
|---|---|
| Magnificent Frigatebird | Sandwich Tern |
| Royal Tern | Elegant Tern |
| Laughing Gull | Black Skimmer |
| Brown Pelican | Neotropic Cormorant |
| Roseate Spoonbill | Wood Stork |
| White Ibis | Yellow-crowned Night-Heron |
| Boat-billed Heron | Great Blue Heron |
| Tricolored Heron | Little Blue Heron |
| Snowy Egret | Great Egret |
| Black-bellied Plover | Willet |
| Whimbrel | Black-necked Stilt |
| Ruddy Turnstone | Spotted Sandpiper |
| Semipalmated Sandpiper | Western Sandpiper |
| Least Sandpiper | |

## Tivives Mangroves

At the small beach town of Tivives there is access to good mangroves. Unlike most of the mangrove areas in the country, a boat is not necessary.

The road to Tivives is well-marked off the highway between Carara and Caldera, the port near Puntarenas. Proceed down the gravel road watching for a side road with a blue metal gate on the left. If you reach a police checkpoint you've gone too far, though there is a small estuary with a few shorebirds on the far (south) end of town. At the end of the side road, climb the fence and go back into the woods which will eventually, depending on water levels, be too muddy to walk. In the dry season, rubber boots are not really necessary. Here Mangrove Warbler is common, and other mangrove specialties including Mangrove Vireo are possible.

## Manuel Antonio National Park

This park is famous for its beautiful beaches, and is very popular with Costa Rica's youth. There is some good habitat too, but it is not necessarily worth the high prices and commotion. The road out to Manuel Antonio from the port town of Quepos is now lined with tourist development; keep in mind if you want to swim that there is no sewage treatment facility for any of those hotels.

Quepos is best reached by the highway that continues beyond Jacó through miles of oil palm plantations. The road is being upgraded and will probably be finished by the time you read this. Quepos is about 3 hours from San José. There is plenty of accommodation in all price ranges in Quepos and Manuel Antonio.

Due to crowd control measures it is a bit hard to get into the park before 8 a.m., though there is a back road with a gate. Park rangers here are among the least friendly anywhere in the system. There is birding along the road to Quepos for species of broken habitats and second growth, but the best habitat is on Punta Catedral and past the first beaches.

There is also a patch of scrubby mangroves north of Quepos that might produce Mangrove Hummingbird and other specialties.

# Birds of Manuel Antonio National Park and vicinity

Brown Booby
Broad-winged Hawk
Gray-headed Chachalaca
Laughing Gull
Sandwich Tern
Pale-vented Pigeon
Blue Ground-Dove
Orange-chinned Parakeet
Red-lored Parrot
Ferruginous Pygmy-Owl
Vaux's Swift
Little Hermit
Charming Hummingbird
Purple-crowned Fairy
Ringed Kingfisher
Golden-naped Woodpecker
Tawny-winged Woodcreeper
Black-hooded Antshrike
Chestnut-backed Antbird
Yellow-olive Flycatcher
Great Crested Flycatcher
White-winged Becard
Golden-collared Manakin
Gray-breasted Martin
Black-bellied Wren
Plain Wren
Tropical Gnatcatcher
Lesser Greenlet
Prothonotary Warbler

Roadside Hawk
Yellow-headed Caracara
Gray-necked Wood-Rail
Royal Tern
Elegant Tern
Scaled Pigeon
White-tipped Dove
White-crowned Parrot
Smooth-billed Ani
Pauraque
Lesser Swallow-tailed Swift
Blue-throated Goldentail
Rufous-tailed Hummingbird
Ruby-throated Hummingbird
Green Kingfisher
Red-crowned Woodpecker
Streak-headed Woodcreeper
Dot-winged Antwren
S. Beardless Tyrannulet
Yellow-bellied Flycatcher
Streaked Flycatcher
Black-crowned Tityra
Red-capped Manakin
S. Rough-winged Swallow
Riverside Wren
Long-billed Gnatwren
Philadelphia Vireo
Tennessee Warbler
Chestnut-sided Warbler

Worm-eating Warbler
Bananaquit
Shining Honeycreeper
Yellow-crowned Euphonia
Buff-throated Saltator
Orange-billed Sparrow
White-collared Seedeater
Eastern Meadowlark
House Sparrow

Kentucky Warbler
Golden-hooded Tanager
Red-legged Honeycreeper
Summer Tanager
Blue-black Grosbeak
Black-striped Sparrow
Variable Seedeater
Northern Oriole

## Corcovado National Park

Corcovado National Park is something of an enigma, as it has some of the best habitat in the country and some of the worst access. Problems with gold miners resulted in the park being temporarily closed to tourists in 1994 and again in 1995. This has happened several times in the past. Presently reservations are required to visit the park, so inquire at the park service office in San José.

You are better off not having a car to worry about while you are in the park. Despite (or because of) the difficult access, birding can be spectacular, with large numbers of Scarlet Macaws, a good chance for Great Curassow, and even the possibility of Baird's Tapir or Jaguar.

The easiest access is to fly into the airstrip at Sirena from Golfito or San José. This is of course expensive. Inquire at any travel agency in San José for prices and options.

There are several ways to hike in, all requiring at least six hours of walking in the heat. The best choice is probably to take a taxi from the town of Puerto Jimenez around the end of the peninsula to the mining town of Carate, and walk up the beach a little over 3 km to the La Leona station. From here you can get an early start to walk the beach. Low tide is a must, as some areas are headlands with inadequate trails above. If you make it past all the river crossings, and are used to the heat, this can be done in six or seven hours. For more details on other walking options, see Joseph Franke's book *Costa Rica's National Parks and Reserves*.

Camping in the chigger-infested clearing at the Sirena station is your most reliable option. There is some possibility of accommodation in the park building and/or meals, if you can arrange it ahead of time with the National Parks office in San José or the local park office in Puerto Jiménez. Be sure radio contact is made to confirm your arrival, or it may come as a surprise to the station you plan to visit.

Sirena has the best facilities though the forest is not as good, being mostly second-growth. Some of the best birds don't seem to mind, however; the area close to the station can be excellent. The San Pedrillo area in the north has the best forest; it is perhaps best visited from one of the lodges on the north side of the park.

## Birds of Corcovado National Park

| | |
|---|---|
| Great Tinamou | Little Tinamou |
| Brown Booby | King Vulture |
| White Hawk | Mangrove Black-Hawk |
| Red-throated Caracara | Laughing Falcon |
| Bat Falcon | Crested Guan |
| Great Curassow | Marbled Wood-Quail |
| White-throated Crake | Gray-necked Wood-Rail |
| Sungrebe | Wilson's Plover |
| Surfbird | Sanderling |
| Least Sandpiper | Laughing Gull |
| Pale-vented Pigeon | Short-billed Pigeon |
| White-tipped Dove | Gray-chested Dove |
| Ruddy Quail-Dove | Crimson-fronted Parakeet |
| Scarlet Macaw | Orange-chinned Parakeet |
| Brown-hooded Parrot | White-crowned Parrot |
| Mealy Parrot | Squirrel Cuckoo |
| Striped Cuckoo | Smooth-billed Ani |
| Spectacled Owl | Mottled Owl |
| Pauraque | Band-rumped Swift |

L. Swallow-tailed Swift
Band-tailed Barbthroat
Little Hermit
White-necked Jacobin
White-crested Coquette
Violet-crowned Woodnymph
Charming Hummingbird
Rufous-tailed Hummingbird
Long-billed Starthroat
Violaceous Trogon
Slaty-tailed Trogon
Ringed Kingfisher
White-necked Puffbird
Rufous-tailed Jacamar
Chestnut-mandibled Toucan
Golden-naped Woodpecker
Pale-billed Woodpecker
Striped Woodhaunter
Scaly-throated Leaftosser
Long-tailed Woodcreeper
Buff-throated Woodcreeper
Streak-headed Woodcreeper
Black-hooded Antshrike
Slaty Antwren
Dusky Antbird
Chestnut-backed Antbird
Black-faced Antthrush
Paltry Tyrannulet
S. Beardless Tyrannulet
Yellow Tyrannulet
Northern Bentbill
Golden-crowned Spadebill

Bronzy Hermit
Long-tailed Hermit
Scaly-breasted Hummingbird
Violet-headed Hummingbird
Garden Emerald
Blue-throated Goldentail
Mangrove Hummingbird
Purple-crowned Fairy
Baird's Trogon
Black-throated Trogon
Blue-crowned Motmot
Am. Pygmy Kingfisher
White-whiskered Puffbird
Fiery-billed Aracari
Olivaceous Piculet
Red-crowned Woodpecker
Slaty Spinetail
Plain Xenops
Tawny-winged Woodcreeper
Wedge-billed Woodcreeper
Black-striped Woodcreeper
Great Antshrike
Plain Antvireo
Dot-winged Antwren
Bare-crowned Antbird
Bicolored Antbird
Spectacled Antpitta
Yellow-bellied Tyrannulet
Ochre-bellied Flycatcher
Scale-crested Pygmy-Tyrant
Eye-ringed Flatbill
Ruddy-tailed Flycatcher

Sulphur-rumped Flycatcher
Rufous Mourner
Cinnamon Becard
Rufous Piha
Yellow-billed Cotinga
Golden-collared Manakin
Red-capped Manakin
Riverside Wren
White-breasted Wood-Wren
Long-billed Gnatwren
Scrub Greenlet
Lesser Greenlet
Tennessee Warbler
Kentucky Warbler
Gray-crowned Yellowthroat
Buff-rumped Warbler
Golden-hooded Tanager
Blue Dacnis
Red-legged Honeycreeper
White-vented Euphonia
Gray-headed Tanager
White-shouldered Tanager
Hepatic Tanager
Buff-throated Saltator
Orange-billed Sparrow
Blue-black Grassquit
Yellow-bellied Seedeater
Northern Oriole
Scarlet-rumped Cacique

Bright-rumped Attila
Dusky-capped Flycatcher
White-winged Becard
Turquoise Cotinga
Thrushlike Manakin
Blue-crowned Manakin
Black-bellied Wren
Plain Wren
S. Nightingale-Wren
Tropical Gnatcatcher
Tawny-crowned Greenlet
Green Shrike-Vireo
Prothonotary Warbler
Mourning Warbler
Canada Warbler
Bananaquit
Scarlet-thighed Dacnis
Shining Honeycreeper
Yellow-crowned Euphonia
Olive-backed Euphonia
White-throated Shrike-Tanager
Black-cheeked Ant-Tanager
Summer Tanager
Blue-black Grosbeak
Black-striped Sparrow
Variable Seedeater
Thick-billed Seed-Finch
Yellow-billed Cacique

## Golfito National Wildlife Refuge

The forest in the hills around the town of Golfito exists as an alternative to birding Corcovado. The habitat above town is good but on steep ground, so birding is restricted to the road. A back road out of Golfito is a much better option, and can be birded without a car.

Golfito is popular with Ticos attracted to the "duty free" zone at the edge to town. In order to buy here one must stay overnight, so there is plenty of accommodation in various price ranges, including several lodges. Prices at the Zona Libre are low by Costa Rican standards but will be of no interest to foreigners.

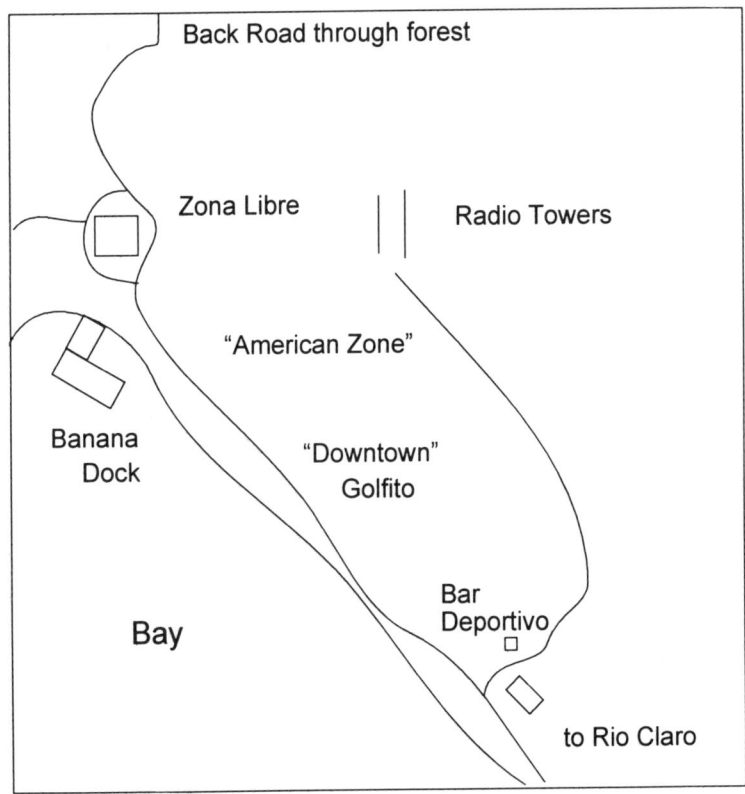

To reach the back road, bear right around the walled Zona Libre, and turn right on a paved road just before a school. This goes through some small housing developments before the road turns to gravel and passes into the wildlife refuge. The Golfito Centro bus line ends about where the road enters the forest.

The road is quite good most of the way through to the Pan-American highway, though several creek crossings require high clearance. The forest is mostly disturbed, but can be good birding with possibilities including Black-cheeked Ant-Tanager. At the far end, it passes through some interesting rice fields with Red-breasted Blackbirds and seedeaters. To reach this area directly from the highway, look for the turn-off to the Eco-Lodge Esquinas at Villa Briceño (Km. 37). The lodge also allows free access to the trail system in the Piedras Blancas park.

The road up to the communications towers above town takes off alongside the first soccer field on the right as you enter Golfito, and passes by the Bar Deportivo. It is steep but should be passible in any season with a 2WD vehicle. Once the road finally levels off the habitat becomes mostly second-growth or active fields. The view over the Golfo Dulce is impressive.

## Birds of Golfito

Great Tinamou
King Vulture
Mangrove Black-Hawk
Bat Falcon
White-throated Crake
Wilson's Plover
Laughing Gull
Short-billed Pigeon
Gray-chested Dove
Crimson-fronted Parakeet
Brown-hooded Parrot
Mealy Parrot
Smooth-billed Ani
Mottled Owl

Little Tinamou
White Hawk
Laughing Falcon
Marbled Wood-Quail
Gray-necked Wood-Rail
Least Sandpiper
Pale-vented Pigeon
White-tipped Dove
Ruddy Quail-Dove
Orange-chinned Parakeet
White-crowned Parrot
Squirrel Cuckoo
Spectacled Owl
Pauraque

| | |
|---|---|
| Band-rumped Swift | L. Swallow-tailed Swift |
| Bronzy Hermit | Band-tailed Barbthroat |
| Long-tailed Hermit | Little Hermit |
| White-necked Jacobin | Violet-headed Hummingbird |
| Garden Emerald | Violet-crowned Woodnymph |
| Blue-throated Goldentail | Charming Hummingbird |
| Mangrove Hummingbird | Rufous-tailed Hummingbird |
| Purple-crowned Fairy | Long-billed Starthroat |
| Baird's Trogon | Violaceous Trogon |
| Black-throated Trogon | Slaty-tailed Trogon |
| Blue-crowned Motmot | Ringed Kingfisher |
| Am. Pygmy Kingfisher | White-necked Puffbird |
| White-whiskered Puffbird | Rufous-tailed Jacamar |
| Fiery-billed Aracari | Chestnut-mandibled Toucan |
| Olivaceous Piculet | Golden-naped Woodpecker |
| Rufous-winged Woodpecker | Red-crowned Woodpecker |
| Pale-billed Woodpecker | Slaty Spinetail |
| Plain Xenops | Scaly-throated Leaftosser |
| Tawny-winged Woodcreeper | Long-tailed Woodcreeper |
| Wedge-billed Woodcreeper | Buff-throated Woodcreeper |
| Black-striped Woodcreeper | Streak-headed Woodcreeper |
| Great Antshrike | Black-hooded Antshrike |
| Plain Antvireo | Dot-winged Antwren |
| Dusky Antbird | Chestnut-backed Antbird |
| Black-faced Antthrush | Spectacled Antpitta |
| Paltry Tyrannulet | Yellow-bellied Tyrannulet |
| S. Beardless Tyrannulet | Ochre-bellied Flycatcher |
| Yellow Tyrannulet | Scale-crested Pygmy-Tyrant |
| Northern Bentbill | Eye-ringed Flatbill |
| Golden-crowned Spadebill | Ruddy-tailed Flycatcher |
| Sulphur-rumped Flycatcher | Bright-rumped Attila |
| Rufous Mourner | Dusky-capped Flycatcher |

| | |
|---|---|
| Cinnamon Becard | White-winged Becard |
| Yellow-billed Cotinga | Thrushlike Manakin |
| Golden-collared Manakin | Blue-crowned Manakin |
| Red-capped Manakin | Black-bellied Wren |
| Riverside Wren | Plain Wren |
| White-breasted Wood-Wren | S. Nightingale Wren |
| Long-billed Gnatwren | Tropical Gnatcatcher |
| Scrub Greenlet | Tawny-crowned Greenlet |
| Lesser Greenlet | Green Shrike-Vireo |
| Tennessee Warbler | Prothonotary Warbler |
| Kentucky Warbler | Mourning Warbler |
| Gray-crowned Yellowthroat | Canada Warbler |
| Buff-rumped Warbler | Bananaquit |
| Golden-hooded Tanager | Scarlet-thighed Dacnis |
| Blue Dacnis | Shining Honeycreeper |
| Red-legged Honeycreeper | Yellow-crowned Euphonia |
| White-vented Euphonia | Spot-crowned Euphonia |
| Gray-headed Tanager | White-throated Shrike-Tanager |
| White-shouldered Tanager | Black-cheeked Ant-Tanager |
| Summer Tanager | Buff-throated Saltator |
| Blue-black Grosbeak | Orange-billed Sparrow |
| Black-striped Sparrow | Blue-black Grassquit |
| Variable Seedeater | Yellow-bellied Seedeater |
| Thick-billed Seed-Finch | Northern Oriole |
| Yellow-billed Cacique | Scarlet-rumped Cacique |

# Guanacaste

# Guanacaste

Costa Rica's northwestern province of Guanacaste provides easy access to the dry forest typical of the Pacific slope of Central America. Most of the province is deforested and full of exotic African savannah grasses, but parks preserve substancial areas of remanent or second-growth forest.

The Tropical Dry Forest has the most pronounced dry season of the various Costa Rican life zones, and many trees drop their leaves during the driest period from December through March. Nonetheless, some of these same trees are flowering, and birding is generally good at that time. The weather may be a problem, as the wind howls down off Lake Nicaragua and can severely reduce productivity by 9 a.m. The national parks don't officially open until 8 a.m., so camping makes good sense.

Birding is relatively easy due to the sparse vegetation and open country, and you can find most of the specialties in a short time. There are a number of nice parks in Guanacaste that are not included here because they are difficult to access, or do not have substantially different birdlife. Guanacaste National Park, across the road from Santa Rosa, and Rincón de la Vieja National Park, above Liberia, are of considerable importance from a conservation standpoint. These are worth a visit if you have time.

Those interested in mammals will find Guanacaste rewarding, since the sparse vegetation makes for easier viewing than elsewhere in the country. The two common species of monkeys (Mantled Howler and White-faced) are relatively easy to find, especially in the dry season. Palo Verde is good for monkeys. Just don't stand underneath them.

The Lomas de Barbudal Biological Reserve mentioned in some guidebooks burned in almost its entirety in March 1994. It unfortunately won't be much of a birding site for years to come.

Much of Costa Rica's mass tourism development is taking place in Guanacaste towards the Nicaraguan border, which may eventually make access more convenient as flights go to the recently-upgraded airport at Liberia. Until then, there is accommodation near the beaches of Playa Hermosa, Playas del Coco, Tamarindo, Sámara, and in the larger towns.

# Santa Rosa National Park

This park preserves history from Costa Rica's less than bellicose past as well as some excellent Tropical Dry Forest. If the road to Playa Naranjo is open and passible, the best place to bird is down near the beaches and around Estero Real. If not, there is plenty of good habitat along the paved road.

The park has been expanded in recent years in an attempt to rehabilitate some of the unproductive cattle country on every side, and has been made part of the Guanacaste Conservation Area in some sort of bureaucratic reshuffling. The best birding is in the Santa Rosa sector.

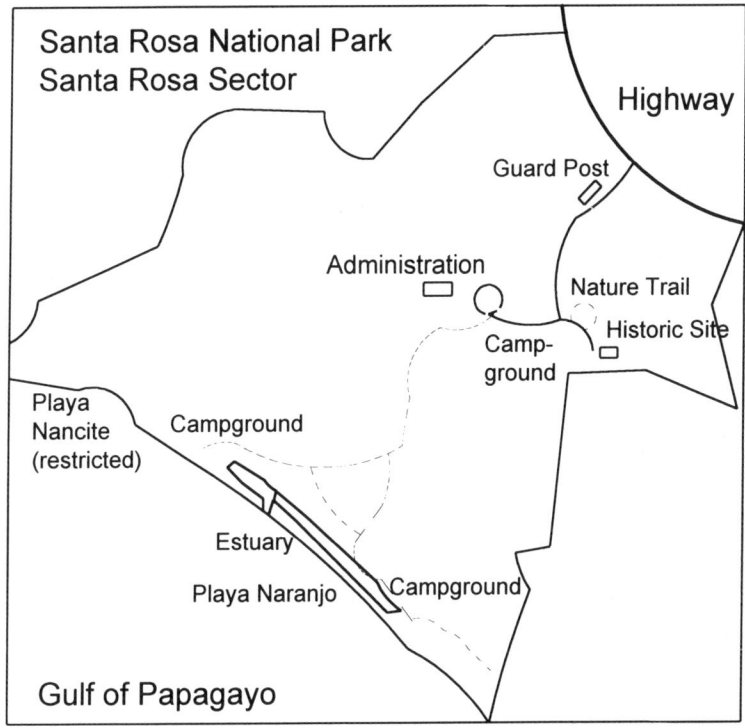

The park is easily reached via Liberia. From town continue 35 km north along the Pan-American highway towards Nicaragua. There are plenty of accommodations in Liberia. At a pleasant campground near the park headquarters, you can listen to Pacific Screech-Owls. There are also campgrounds at Playa Naranjo, but these can be full of noisy surfers when the road is open. It's a 14 km walk during the rainy season.

Some of the best birding in the park is in the bottomlands near Playa Naranjo. The road down to the beach requires 4WD, and is only open when dry. The main attraction here is the Great Curassow, which has a reasonable population in the park. To find the curassow, walk the flat part of the road to Playa Nancite listening for the rustling of dead leaves in the underbrush. This is considerably easier before the lizards are out. Crested Guan (arboreal) is also in the area.

## Birds of Santa Rosa National Park

| | |
|---|---|
| Thicket Tinamou | Brown Booby |
| Brown Pelican | Magnificent Frigatebird |
| King Vulture | White-tailed Kite |
| Double-toothed Kite | Plumbeous Kite |
| Mangrove Black-Hawk | Great Black-Hawk |
| White-tailed Hawk | Zone-tailed Hawk |
| Harris' Hawk | Roadside Hawk |
| Gray Hawk | Osprey |
| Crested Caracara | Laughing Falcon |
| American Kestrel | Plain Chachalaca |
| Crested Guan | Great Curassow |
| Spot-bellied Bobwhite | Gray-necked Wood-Rail |
| Black-bellied Plover | Willet |
| Whimbrel | Sanderling |
| Laughing Gull | Royal Tern |
| Red-billed Pigeon | White-winged Dove |
| Inca Dove | Common Ground-Dove |
| Blue Ground-Dove | White-tipped Dove |
| Orange-fronted Parakeet | Orange-chinned Parakeet |
| White-fronted Parrot | Yellow-naped Parrot |
| Mangrove Cuckoo | Lesser Ground-Cuckoo |
| Squirrel Cuckoo | Groove-billed Ani |

Barn Owl
Ferruginous Pygmy-Owl
Lesser Nighthawk
Green-breasted Mango
Blue-throated Goldentail
Steely-vented Hummingbird
Plain-capped Starthroat
Black-headed Trogon
Violaceous Trogon
Blue-crowned Motmot
Collared Aracari
Hoffman's Woodpecker
Pale-billed Woodpecker
Ivory-billed Woodcreeper
Barred Antshrike
Northern Bentbill
Greenish Elaenia
Yellow-olive Flycatcher
Bright-rumped Attila
Nutting's Flycatcher
Brown-crested Flycatcher
Boat-billed Flycatcher
Streaked Flycatcher
Piratic Flycatcher
Western Kingbird
Rose-throated Becard
Black-crowned Tityra
Gray-breasted Martin
Barn Swallow
Rufous-naped Wren
Banded Wren
House Wren
Clay-colored Thrush

Pacific Screech-Owl
Mottled Owl
Pauraque
Fork-tailed Emerald
Cinnamon Hummingbird
Rufous-tailed Hummingbird
Ruby-throated Hummingbird
Elegant Trogon
Turquoise-browed Motmot
White-necked Puffbird
Keel-billed Toucan
Lineated Woodpecker
Ruddy Woodcreeper
Streak-headed Woodcreeper
N. Beardless Tyrannulet
Yellow-bellied Elaenia
Common Tody-Flycatcher
Stub-tailed Spadebill
Dusky-capped Flycatcher
Great Crested Flycatcher
Great Kiskadee
Social Flycatcher
Sulphur-bellied Flycatcher
Tropical Kingbird
Scissor-tailed Flycatcher
Masked Tityra
Long-tailed Manakin
Mangrove Swallow
White-throated Magpie-Jay
Rufous-and-White Wren
Plain Wren
White-lored Gnatcatcher
Swainson's Thrush

| | |
|---|---|
| Yellow-green Vireo | Lesser Greenlet |
| Tennessee Warbler | Yellow Warbler |
| Northern Waterthrush | Gray-crowned Yellowthroat |
| Rufous-capped Warbler | Red-legged Honeycreeper |
| Thick-billed Euphonia | Scrub Euphonia |
| Blue-gray Tanager | Summer Tanager |
| Western Tanager | Grayish Saltator |
| Buff-throated Saltator | Rose-breasted Grosbeak |
| Indigo Bunting | Olive Sparrow |
| White-collared Seedeater | Yellow-faced Grassquit |
| Northern Oriole | Streak-backed Oriole |
| Spot-breasted Oriole | |

## Palo Verde National Park

The marshes at the mouth of the Río Tempisque are the main attraction here, though they are difficult to access. This is one of the last strongholds of the Jabiru in Central America; it can be found with some regularity. There is good Tropical Dry Forest as well, and an impressive number of large iguanas. Africanized bees can be a problem.

The park is reached by a rough road from the town of Bagaces on the Pan-American highway. The road should be passible year-round. If you get a flat tire, return to town and get it fixed before proceeding, as there is little traffic to take you out if you get another. The nearest hotels are in Liberia, though you can camp or stay at the OTS station (see p. 79 for information).

The marshes hold large numbers of water birds in the November to March dry season, though some leave as water becomes scarce after January. Mostly the regular species are the same as in Florida, but numbers can be impressive. The best place to view the marshes is between the dock and the OTS station. Otherwise there are a number of trails that lead into the forest.

Palo Verde is a good place to see monkeys, with troops of Black (Mantled) Howlers and White-faced Capuchins easy to find. The latter especially like the mango trees near park HQ, when there's a crop.

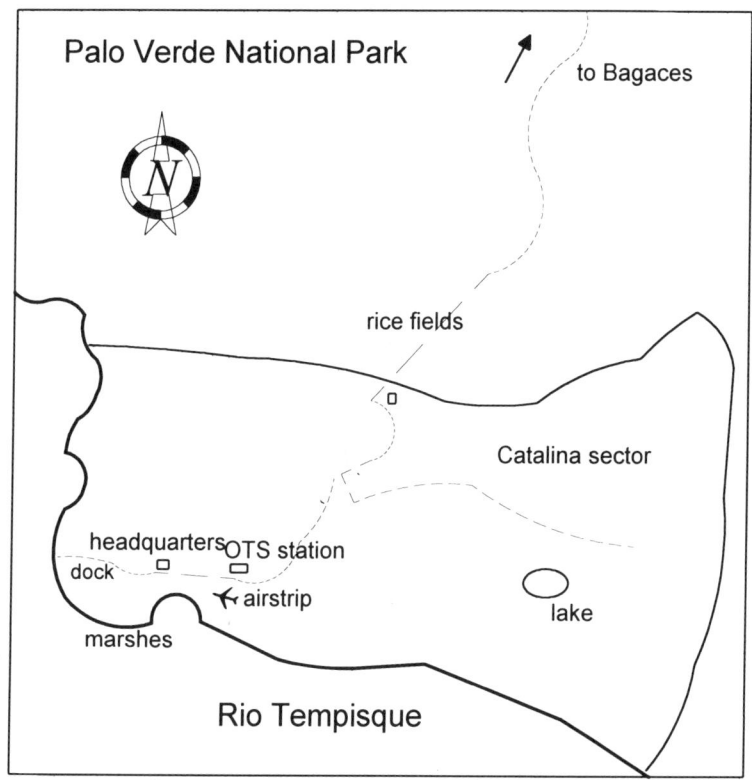

## Birds of Palo Verde National Park

Thicket Tinamou
Pied-billed Grebe
Neotropic Cormorant
Great Blue Heron
Snowy Egret
Black-crowned Night-Heron
Boat-billed Heron
Glossy Ibis
Jabiru
Black-bellied Whistling-Duck

Least Grebe
Anhinga
Bare-throated Tiger-Heron
Great Egret
Green Heron
Yellow-crowned Night-Heron
White Ibis
Roseate Spoonbill
Wood Stork
Fulvous Whistling-Duck

| | |
|---|---|
| Muscovy | Northern Pintail |
| Blue-winged Teal | Northern Shoveler |
| American Wigeon | Masked Duck |
| Lesser Yellow-headed Vulture | King Vulture |
| White-tailed Kite | Snail Kite |
| Double-toothed Kite | Plumbeous Kite |
| Mangrove Black-Hawk | Great Black-Hawk |
| White-tailed Hawk | Zone-tailed Hawk |
| Harris' Hawk | Roadside Hawk |
| Gray Hawk | Osprey |
| Crested Caracara | Laughing Falcon |
| American Kestrel | Plain Chachalaca |
| Spot-bellied Bobwhite | White-throated Crake |
| Gray-necked Wood-Rail | Purple Gallinule |
| Common Moorhen | American Coot |
| Limpkin | Black-bellied Plover |
| Willet | Black Tern |
| Red-billed Pigeon | White-winged Dove |
| Inca Dove | Common Ground-Dove |
| Blue Ground-Dove | White-tipped Dove |
| Orange-fronted Parakeet | Orange-chinned Parakeet |
| White-fronted Parrot | Yellow-naped Parrot |
| Mangrove Cuckoo | Lesser Ground-Cuckoo |
| Squirrel Cuckoo | Groove-billed Ani |
| Barn Owl | Pacific Screech-Owl |
| Ferruginous Pygmy-Owl | Mottled Owl |
| Lesser Nighthawk | Pauraque |
| Green-breasted Mango | Fork-tailed Emerald |
| Cinnamon Hummingbird | Steely-vented Hummingbird |
| Rufous-tailed Hummingbird | Plain-capped Starthroat |
| Ruby-throated Hummingbird | Black-headed Trogon |
| Elegant Trogon | Violaceous Trogon |
| Turquoise-browed Motmot | Collared Aracari |

Keel-billed Toucan
Lineated Woodpecker
Ivory-billed Woodcreeper
Barred Antshrike
Yellow-bellied Elaenia
Common Tody-Flycatcher
Stub-tailed Spadebill
Dusky-capped Flycatcher
Great Crested Flycatcher
Great Kiskadee
Social Flycatcher
Sulphur-bellied Flycatcher
Tropical Kingbird
Rose-throated Becard
Long-tailed Manakin
Mangrove Swallow
White-throated Magpie-Jay
Rufous-and-White Wren
Plain Wren
White-lored Gnatcatcher
Swainson's Thrush
Lesser Greenlet
Yellow Warbler
Gray-crowned Yellowthroat
Red-legged Honeycreeper
Scrub Euphonia
Summer Tanager
Grayish Saltator
Rose-breasted Grosbeak
Olive Sparrow
Yellow-faced Grassquit
Streak-backed Oriole

Hoffman's Woodpecker
Pale-billed Woodpecker
Streak-headed Woodcreeper
N. Beardless Tyrannulet
Greenish Elaenia
Yellow-olive Flycatcher
Bright-rumped Attila
Nutting's Flycatcher
Brown-crested Flycatcher
Boat-billed Flycatcher
Streaked Flycatcher
Piratic Flycatcher
Scissor-tailed Flycatcher
Masked Tityra
Gray-breasted Martin
Barn Swallow
Rufous-naped Wren
Banded Wren
House Wren
Clay-colored Thrush
Yellow-green Vireo
Tennessee Warbler
Northern Waterthrush
Rufous-capped Warbler
Thick-billed Euphonia
Blue-gray Tanager
Western Tanager
Buff-throated Saltator
Indigo Bunting
White-collared Seedeater
Northern Oriole

## Barra Honda National Park

This park exists to control and protect some fine limestone caves, but also has some interesting, if sparse, Tropical Dry Forest. You would probably need a guide to see the caves, and will likely be approached by one. Howler monkeys are particularly easy to see here.

Access is via the Tempisque ferry and the highway to Nicoya, with a well-signed road leading off to the right about 6 km west of the ferry. There are signs at the intersections. A small stream crosses the road near the park so you would likely need 4WD during the rainy season.

Ask at the park entrance for permission to go in early; the only trails go right by the building, so it would be difficult to sneak in. Birding is good along any of the trails, and the relatively sparse vegetation makes secretive species like Thicket Tinamou, Lesser Ground Cuckoo, and Long-tailed Manakin easier to observe here than elsewhere.

## Birds of Barra Honda National Park

Thicket Tinamou
Gray Hawk
Plain Chachalaca
White-fronted Parrot
Lesser Ground-Cuckoo
Pacific Screech-Owl
Lesser Nighthawk
Steely-vented Hummingbird
Cinnamon Hummingbird
Black-headed Trogon
Turquoise-browed Motmot
Pale-billed Woodpecker
Streaked Flycatcher
Brown-crested Flycatcher

Roadside Hawk
Broad-winged Hawk
Orange-fronted Parakeet
Inca Dove
Mangrove Cuckoo
Pauraque
Fork-tailed Emerald
Ruby-throated Hummingbird
Elegant Trogon
Blue-crowned Motmot
Hoffman's Woodpecker
Yellow-olive Flycatcher
Nutting's Flycatcher
Great Crested Flycatcher

| | |
|---|---|
| Long-tailed Manakin | Rufous-naped Wren |
| Banded Wren | White-lored Gnatcatcher |
| Yellow-throated Vireo | Yellow-green Vireo |
| Lesser Greenlet | Tennessee Warbler |
| Rufous-capped Warbler | Olive Sparrow |

## Beaches

It is possible to bird Guanacaste from one of the beach resorts that are springing up on the outer coast of the Nicoya peninsula. Habitat around the towns is not as good as in the national parks but the common species can be seen.

The best site with much infrastructure is probably Tamarindo, which in addition to dry forest and scrub has an estuary with mangroves. You can rent a boat in town to visit the mangroves, though most of the Pacific coast mangrove specialties don't range this far north. The estuary also has a few shorebirds.

Other areas that have at least some habitat and have a decent range of accommodation include Playa Hermosa and Playas del Coco near Liberia, and the Nosara/Samara area further south.

At the southern tip of the Nicoya peninsula, Montezuma and the Cabo Blanco Strict Nature Preserve have good scenery and some interesting birding. The forest on the southern end of the Nicoya peninsula is slightly more moist than Guanacaste, though the birdlife is generally similar.

# The Mountains

# The Mountains

The volcanic ranges that form the spine of Costa Rica are its most striking geographical feature, and the source of most of its biological diversity. Many of the best areas are within easy reach of San José, so it is possible to visit several sites without changing hotels.

A number of good areas are not covered here due to access problems. If you don't mind bad roads, Rincón de la Vieja National Park, on the volcano of the same name above Liberia, is well worth visiting. The area around Volcán Arenal is popular with tourists wanting to see the active volcano, but there is not too much habitat for birds. Finally, a vast area of the Cordillera de Talamanca is protected by Chirripó and La Amistad National Parks, but the habitat types found there are much more easily visited at Cerro de la Muerte.

The locations below are biased towards the Caribbean slope. That's mostly where the parks are, but the middle-elevation Caribbean slope does have a very distinctive avifauna, with a number of difficult-to-find species. It is well worth extra time. Most of the Pacific slope near the Valle Central has been deforested, so that zone isn't too productive.

## Braulio Carrillo National Park

Long popular with birders for its good habitat within easy reach of San José, this park has suffered in recent years from a spate of car break-ins and armed robbery. For this reason the well-known Botarrama and Carrillo trails are officially off limits, though a newer trail by the lower park station partially replaces them for birding purposes. Tourists with expensive cameras and binoculars are the main targets. Rental cars are stolen as well. The perpetrators of these crimes are quite nasty by Costa Rican standards; tourists have been tied up and beaten in addition to losing property.

Braulio Carrillo provides scenery unsurpassed in Costa Rica and a good transect of Caribbean slope forest. The main access is along the Guápiles highway, which starts out as Calle 3 northbound in San José. The first park station is about half an hour from the city, the second about 20 minutes further. Due to problems mentioned above, the only official access is now a trail at the lower (second) park station. The trailheads are also readily reached by public transport, the best choice being the Guápiles bus. Ask beforehand if the driver knows where to stop.

The trail by the first park station, just before the Zurquí tunnel, is currently not maintained due to park budgetary constraints and damage from heavy rains in February 1996. It was never very good for birding anyway.

At the other station by the Río Quebrada Gonzales, a loop trail goes into good foothill forest with a mix of middle- and low-elevation species. The station is not really obvious from the highway, so watch for the river; the entrance is to the right just before the bridge (coming from San José). Local species including Yellow-eared Toucanet, Lattice-tailed Trogon, Purplish-backed Quail-Dove, and Ashy-throated Bush-Tanager can be found here. Recent sightings also include Keel-billed Motmot and Black-crowned Antpitta (regular but elusive).

It is also possible to visit another corner of the park, above the town of Sacramento de Heredia. This requires a 4 km uphill hike through pasture along an undrivable road just to get to the park boundary, then another few kilometers through good forest to a lake on Volcán Barva.

## Birds of the Quebrada Gonzalez trail

Great Tinamou
Crested Guan
Short-billed Pigeon
Purplish-backed Quail-Dove
Brown-hooded Parrot
Squirrel Cuckoo
Gray-rumped Swift
Little Hermit
Violet-headed Hummingbird
Snowcap
Lattice-tailed Trogon
Violaceous Trogon
Collared Aracari
Chestnut-mandibled Toucan
Rufous-winged Woodpecker
Cinnamon Woodpecker

White Hawk
Rufous-fronted Wood-Quail
Gray-chested Dove
Crimson-fronted Parakeet
White-crowned Parrot
White-collared Swift
Green Hermit
White-tipped Sicklebill
Violet-crowned Woodnymph
Purple-crowned Fairy
Black-throated Trogon
Broad-billed Motmot
Yellow-eared Toucan
Black-cheeked W
Smoky-brown W
Pale-billed W

| | |
|---|---|
| Spotted Woodcreeper | Wedge-billed Woodcreeper |
| Brown-billed Scythebill | Spotted Barbtail |
| Plain Xenops | Russet Antshrike |
| Streak-crowned Antvireo | Chestnut-backed Antbird |
| Bicolored Antbird | Spectacled Antpitta |
| White-crowned Manakin | White-ruffed Manakin |
| Yellowish Flycatcher | Tufted Flycatcher |
| Yellow-olive Flycatcher | Yellow-margined Flycatcher |
| Eye-ringed Flatbill | Rufous-browed Tyrannulet |
| Golden-crowned Spadebill | Olive-striped Flycatcher |
| Ochre-bellied Flycatcher | Band-backed Wren |
| Stripe-breasted Wren | Bay Wren |
| White-breasted Wood-Wren | Pale-vented Thrush |
| Wood Thrush | Swainson's Thrush |
| Black-headed Nightingale-Thrush | Tropical Gnatcatcher |
| Tawny-faced Gnatwren | Green Shrike-Vireo |
| Lesser Greenlet | Bananaquit |
| Tropical Parula | Chestnut-sided Warbler |
| Buff-rumped Warbler | Green Honeycreeper |
| Shining Honeycreeper | Scarlet-thighed Dacnis |
| Emerald Tanager | Silver-throated Tanager |
| Speckled Tanager | Bay-headed Tanager |
| Tawny-capped Euphonia | Olive-backed Euphonia |
| Blue-and-Gold Tanager | Olive Tanager |
| Tawny-crested Tanager | White-shouldered Tanager |
| Summer Tanager | Dusky-faced Tanager |
| Scarlet-rumped Tanager | Ashy-throated Bush-Tanager |
| Black-and-Yellow Tanager | Slate-colored Grosbeak |
| Orange-billed Sparrow | Black-faced Grosbeak |
| Scarlet-rumped Cacique | Montezuma Oropendola |

## Tapantí National Park

This park preserves a watershed with superb Caribbean foothill forest. Access is easy with a car, very difficult by public transport. Park personnel are familiar with birders, so just drive by the station if you arrive before 8 a.m. You can pay on the way out.

From San José, pass through Cartago to the town of Paraíso. The highway forces you into the center of town, where you make a right on Calle 2 towards the town of Orosi. Continue on the paved road through Orosi to an electrical substation, where the road goes straight and becomes broken asphalt, potholes, and mud if it has rained. The road is nonetheless passible with an ordinary car at any time.

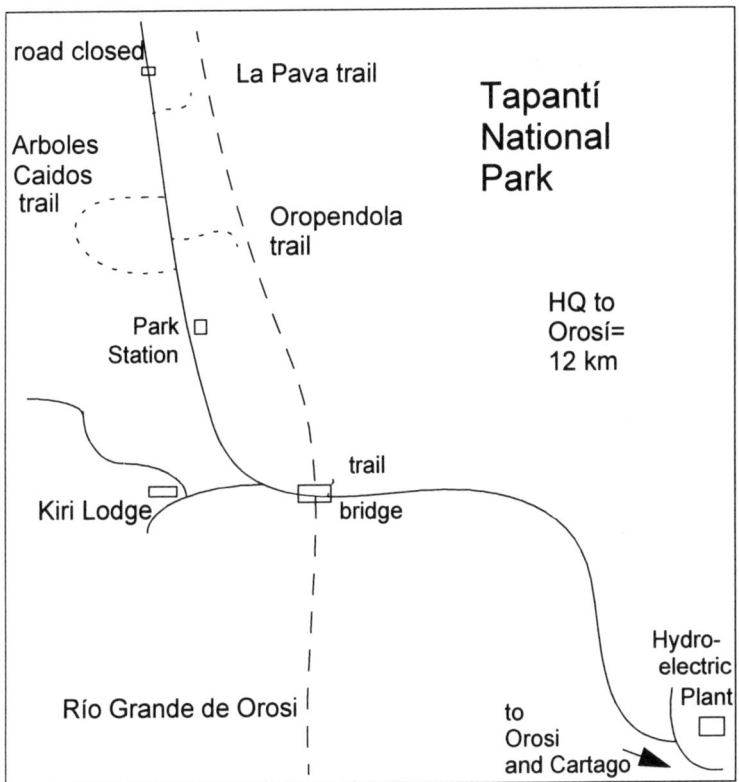

About 7 km from Orosi, there is a river crossing with good birding early in the morning. After a couple more kilometers, the park station is on the right. From San José to the park is about a 1½ hour drive. Note: as of 1996 there were several culverts washed out on the far side of Orosi, necessitating a small detour.This might be fixed by the time you arrive; but since it's already been two years it might not be.

The best birding is along a rather good gravel road through the park that eventually leads to a hydroelectric plant on the mountain. The road is closed to vehicle traffic after about 5 km, but you can and should walk it further. The several trails are good, but less important. There is some general use, but during the week you should have the place to yourself.

Tapantí is readily reached as a day trip from San José but there are also a number of modest hotels in the Orosí valley. Right by the entrance is the Kiri Lodge. There is a restaurant on the premises as well.

## Birds of Tapantí

| | |
|---|---|
| Black Guan | Black-chested Hawk |
| Band-tailed Pigeon | Ruddy Pigeon |
| Brown-hooded Parrot | White-crowned Parrot |
| Vaux's Swift | Green Hermit |
| Green-crowned Brilliant | Green-fronted Lancebill |
| Black-bellied Hummingbird | White-bellied Mountain-gem |
| Variable Mountain-gem | Green Thorntail |
| Collared Trogon | Emerald Toucanet |
| Red-headed Barbet | Prong-billed Barbet |
| Red-faced Spinetail | Spotted Barbtail |
| Streaked Xenops | Buffy Tuftedcheek |
| Slaty Antwren | Paltry Tyrannulet |
| Olive-striped Flycatcher | Dark Pewee |
| Tufted Flycatcher | Black Phoebe |
| Golden-bellied Flycatcher | Bright-rumped Attila |
| Gray-breasted Wood-Wren | Ochraceous Wren |
| Black-faced Solitaire | Orange-billed Nightingale-Thrush |
| N. Am. Dipper | Brown-capped Vireo |

Rufous-browed Peppershrike
Golden-winged Warbler
Black-throated Green Warbler
Black-and-White Warbler
Wilson's Warbler
Collared Whitestart
Three-striped Warbler
Silver-throated Tanager
Golden-hooded Tanager
Rufous-winged Tanager
Common Bush-Tanager
Tawny-capped Euphonia
Scarlet-rumped Tanager
Sooty-faced Finch
Yellow-faced Grassquit
Rufous-collared Sparrow

Tennessee Warbler
Blackburnian Warbler
Chestnut-sided Warbler
Mourning Warbler
Slate-throated Whitestart
Tropical Parula
Scarlet-thighed Dacnis
Spangle-cheeked Tanager
Bay-headed Tanager
White-winged Tanager
Golden-browed Chlorophonia
Blue-hooded Euphonia
Crimson-collared Tanager
Yellow-throated Brush-Finch
Variable Seedeater
Lesser Goldfinch

## Volcán Irazú National Park

The primary attraction here is a barren volcanic crater at the top of a rather large active volcano, with views to both coasts on a rare clear day. It is easily reached from San José via Cartago.

To reach the park from San José, take the freeway to Cartago. Follow the signs saying "Volcan Irazu" and exit left across the rumble strips. Take the first left, just after a body shop on the left and a hardware store on the right. It is than a long climb up past the towns of Cot and Tierra Blanca, set amidst the main vegetable-growing area of the country. Generally the route is well-signed.

For slightly lower-elevation birding, turn left on a paved road, signed for Prussia, and bird the roadside for a few kilometers until it dead-ends at a park. The forest at the far end of the road is mostly exotic conifers, and doesn't support many birds. Along the road are many Green Violetears, etc. This area can be good on a cold day while you wait for things to warm up a bit at the summit.

The vegetation at the top has not had much time to grow back from the most recent eruption (1963), but supports lots of Volcano Hummingbirds and a few other species, mostly Central American endemics. The parking lot usually has Volcano Junco. If you are not interested in the crater itself, the junco can sometimes be seen (without paying the entrance fee) along the road mentioned below. Look especially where there is bare ash along the cut-banks.

Just below the park gate, a road goes off to the right and down the north side of the mountain into interesting habitat. The road is steep in places but the first part should be passable with an ordinary car.

## Birds of Irazú National Park & vicinity

Buffy-crowned Wood-Partridge
Band-tailed Pigeon
White-collared Swift
Fiery-throated Hummingbird
Volcano Hummingbird
Mountain Elaenia
Blue-and-White Swallow
Sooty Thrush
Flame-throated Warbler
Slate-throated Whitestart
Sooty-capped Bush-Tanager
Yellow-thighed Finch
Volcano Junco

Red-billed Pigeon
Mourning Dove
Green Violetear
Scintillant Hummingbird
Acorn Woodpecker
Black-capped Flycatcher
Black-billed Nightingale-Thrush
Long-tailed Silky-Flycatcher
Wilson's Warbler
Flame-colored Tanager
Large-footed Finch
Rufous-collared Sparrow
Peg-billed Finch

## Guayabo National Monument

This park preserves the most significant pre-Colombian archeological site in Costa Rica. Though not up to the standards of Mesoamerican civilizations further north, it is worth a visit. Most of the forest is only about 35 years old, but the area always seems to have an impressive variety of wintering birds.

To reach the monument, proceed to the town of Turrialba. A couple of blocks after entering the town, a sign directs you to turn left for Santa Rosa and Guayabo. The bridge you want is actually a block back to the left from the street you are now on, so jog left on the last possibility, and then right again, to cross a one-lane bridge that looks like a railroad bridge. Don't go towards Santa Rosa after the initial turn. Once across the bridge, stay on the pavement until a sign directs you up a gravel road to the monument.

There is also a shorter route from Cartago and Volcán Irazú, via Pacayas and Santa Cruz. Unfortunately, the signs for the monument at the turn in Santa Cruz face downhill, so you may have to retrace your path if you leave the town without seeing the signs. This back road into the monument is rough, especially in the rainy season.

There is camping at the monument but it often rains. Nearby on the entrance road, there is a small hotel with a restaurant. There is also a soda (closed Saturdays) within walking distance of the reserve.

The trails through the second-growth around the archeological area are sometimes productive, but the best birding is generally along the road. Usually there are flocks around the campground as well.

## Birds of Guayabo National Monument

| | |
|---|---|
| Little Tinamou | Cattle Egret |
| Turkey Vulture | Black Vulture |
| Double-toothed Kite | Black Hawk-Eagle |
| Gray-headed Chachalaca | Red-billed Pigeon |
| Short-billed Pigeon | Ruddy Ground-Dove |
| White-tipped Dove | Crimson-fronted Parakeet |
| White-crowned Parrot | Squirrel Cuckoo |
| Groove-billed Ani | Pauraque |
| White-collared Swift | Vaux's Swift |
| Green Hermit | Violet Sabrewing |
| Green-crowned Brilliant | Green Violetear |
| Rufous-tailed Hummingbird | Violet-crowned Woodnymph |
| Purple-crowned Fairy | Violaceous Trogon |

| | |
|---|---|
| Broad-billed Motmot | Rufous Motmot |
| Blue-crowned Motmot | Red-headed Barbet |
| Keel-billed Toucan | Collared Aracari |
| Emerald Toucanet | Black-cheeked Woodpecker |
| Golden-olive Woodpecker | Rufous-winged Woodpecker |
| Pale-billed Woodpecker | Lineated Woodpecker |
| Olivaceous Woodcreeper | Streak-headed Woodcreeper |
| Buff-throated Woodcreeper | Spectacled Foliage-gleaner |
| Plain Xenops | Barred Antshrike |
| Slaty Antwren | Black-faced Antthrush |
| White-ruffed Manakin | Cinnamon Becard |
| Rufous Mourner | Masked Tityra |
| Black-crowned Tityra | Tropical Kingbird |
| Great Kiskadee | Social Flycatcher |
| Piratic Flycatcher | Dusky-capped Flycatcher |
| Tropical Pewee | Yellowish Flycatcher |
| Slaty-capped Flycatcher | Eye-ringed Flatbill |
| Yellow-olive Flycatcher | Paltry Tyrannulet |
| Olive-striped Flycatcher | Ochre-bellied Flycatcher |
| Blue-and-White Swallow | Brown Jay |
| Band-backed Wren | Stripe-breasted Wren |
| Plain Wren | Bay Wren |
| House Wren | Gray-breasted Wood-Wren |
| Clay-colored Thrush | Tropical Gnatcatcher |
| Yellow-throated Vireo | Yellow-green Vireo |
| Brown-capped Vireo | Lesser Greenlet |
| Bananaquit | Scarlet-thighed Dacnis |
| Green Honeycreeper | Black-and-White Warbler |
| Golden-winged Warbler | Tennessee Warbler |
| Tropical Parula | Blackburnian Warbler |
| Chestnut-sided Warbler | Kentucky Warbler |
| Mourning Warbler | Wilson's Warbler |
| Slate-throated Whitestart | Golden-crowned Warbler |

Rufous-capped Warbler
Montezuma Oropendola
Tawny-capped Euphonia
Silver-throated Tanager
Bay-headed Tanager
Palm Tanager
Crimson-collared Tanager
Flame-colored Tanager
Black-headed Saltator
Rose-breasted Grosbeak
Yellow-throated Brush-Finch
Rufous-collared Sparrow

Buff-rumped Warbler
Giant Cowbird
Olive-backed Euphonia
Golden-masked Tanager
Blue-gray Tanager
Scarlet-rumped Tanager
Summer Tanager
White-shouldered Tanager
Buff-throated Saltator
Yellow-faced Grassquit
Orange-billed Sparrow

## Volcán Poás National Park

This popular national park offers both an active volcano and some good high-elevation habitat. It is well-visited by both Ticos and foreign tourists, so arrive early and avoid weekends.

The park is readily reached via Alajuela or Heredia. Generally it is easier to go through Heredia. Follow the directions on page 22 to get on the main highway from San José. Upon entering Heredia, you will pass a gas station on the left and then the National University on the right. A few blocks further this street ends: take a left following a Britt Coffee tour sign. After a few more blocks, take a right at another Britt sign and you're on the highway to Sarapiquí. At Vara Blanca go straight towards the national park when the main highway goes down the hill towards Virgen del Socorro and Puerto Viejo.

If you arrive before the gate opens in the morning, there is plenty of habitat by the main road, and also along a dirt road that goes off to the left (just below the gate) towards a small tourist development. If it is cloudy, and it often is, it's probably not worth going into the park itself, since the crater will be fogged in. The best birding is generally before the entrance booth. As with Volcán Irazú, there are relatively few species here, but they are mostly Costa Rica/Chiriquí specialties.

## Birds of Volcán Poás National Park

| | |
|---|---|
| Swallow-tailed Kite | Red-tailed Hawk |
| Barred Hawk | Black Guan |
| Spotted Wood-Quail | Band-tailed Pigeon |
| Barred Parakeet | Bare-shanked Screech-Owl |
| Dusky Nightjar | White-collared Swift |
| Vaux's Swift | Green Violetear |
| Fiery-throated Hummingbird | Magnificent Hummingbird |
| Volcano Hummingbird | Scintillant Hummingbird |
| Resplendent Quetzal | Collared Trogon |
| Emerald Toucanet | Acorn Woodpecker |
| Hairy Woodpecker | Spot-crowned Woodcreeper |
| Ruddy Treerunner | Buffy Tuftedcheek |
| Silvery-fronted Tapaculo | Black-capped Flycatcher |
| Mountain Elaenia | Blue-and-White Swallow |
| Ochraceous Wren | Gray-breasted Wood-Wren |
| Sooty Thrush | Mountain Thrush |
| Ruddy-capped Nightingale-Thrush | Black-billed Nightingale-Thrush |
| Long-tailed Silky-Flycatcher | Black-and-Yellow Silky-Flycatcher |
| Yellow-winged Vireo | Brown-capped Vireo |
| Slaty Flower-piercer | Flame-throated Warbler |
| Black-throated Green Warbler | Wilson's Warbler |
| Collared Whitestart | Black-cheeked Warbler |
| Wrenthrush | Sooty-capped Bush-Tanager |
| Large-footed Finch | Sooty-faced Finch |
| Yellow-thighed Finch | Rufous-collared Sparrow |

## Virgen del Socorro

Good lower-elevation foothill forest can be visited from the Sarapiquí highway at the Virgen del Socorro forest reserve. Birding is from a rough dirt road that crosses the Sarapiquí river and doubles back up the other side of the canyon, to some farms on the opposite ridge.

The entrance to the forest reserve is about 4 km below the El Angel marmalade factory below Varablanca, which is about 1½ hours from San José via Heredia. The road is the first turn on the right after the factory, and requires a sharp right turn. If driving a 4WD vehicle, be careful not to heat up the brakes on the long curving decent from the pass.

Birding is good along the road in both directions from the bridge. A good trail with some additional species goes upstream along the river about 50 feet from the bridge, on the highway side. Conditions are more difficult along the trail due to the closed canopy and water noise.

## Birds of Virgen del Socorro

Swallow-tailed Kite
Broad-winged Hawk
Crimson-fronted Parakeet
Squirrel Cuckoo
Vaux's Swift
Little Hermit
Violet-headed Hummingbird
Coppery-headed Emerald
Green-crowned Brilliant
Collared Trogon
Red-headed Barbet
Smoky-brown Woodpecker
Spotted Woodcreeper
Red-faced Spinetail
Immaculate Antbird
Paltry Tyrannulet
Rufous-browed Tyrannulet
Scale-crested Pygmy-Tyrant
Tufted Flycatcher
Golden-bellied Flycatcher
Azure-hooded Jay
Northern Nightingale Wren

Black-chested Hawk
Bat Falcon
White-crowned Parrot
White-collared Swift
Green Hermit
Brown Violetear
Violet-crowned Woodnymph
Green Thorntail
Purple-crowned Fairy
Collared Aracari
Prong-billed Barbet
Rufous-winged Woodpecker
Streak-headed Woodcreeper
Spotted Barbtail
Slaty Antwren
Torrent Tyrannulet
Slaty-capped Flycatcher
Yellow-margined Flycatcher
Yellow-bellied Flycatcher
White-ruffed Manakin
Bay Wren
N. Am. Dipper

| | |
|---|---|
| Tawny-faced Gnatwren | Slaty-backed Nightingale-Thrush |
| Wood Thrush | Pale-vented Thrush |
| Yellow-throated Vireo | Lesser Greenlet |
| Tropical Parula | Golden-winged Warbler |
| Black-throated Green Warbler | Black-and-White Warbler |
| Chestnut-sided Warbler | Blackburnian Warbler |
| Kentucky Warbler | Mourning Warbler |
| Wilson's Warbler | Slate-throated Whitestart |
| Three-striped Warbler | Golden-crowned Warbler |
| Bananaquit | Emerald Tanager |
| Silver-throated Tanager | Bay-headed Tanager |
| Speckled Tanager | Scarlet-thighed Dacnis |
| Tawny-capped Euphonia | Crimson-collared Tanager |
| Scarlet-rumped Tanager | Common Bush-Tanager |
| Black-and-Yellow Tanager | Slate-colored Grosbeak |
| Rose-breasted Grosbeak | Sooty-faced Finch |

## Cerro de la Muerte

Crossing the Cerro de la Muerte, the Pan-American Highway reaches its highest point. The highway provides easy access to a range of habitats, including a type of shrubby brushland vaguely similar to the Andean páramo, and high-elevation oak forest. Weather is often a problem here, even in the dry season. Birding can still be good in a light rain or fog, but much wind can ruin your day.

This area is a good choice for high-elevation species, particularly if you don't want to go to Monteverde. Also available here is the Resplendant Quetzal, which has become something of a cottage industry on the Cerro.

Access is easy from San José, as all sites are near the highway, and the far end of the birding area can be reached in 100 minutes of cautious driving. Slides and sunken grades reduce the road to one lane in a number of places. The highway is marked with concrete kilometer posts painted yellow; enough of these are still in view to provide a ready reference to your position. These will be referred to regularly in the following account.

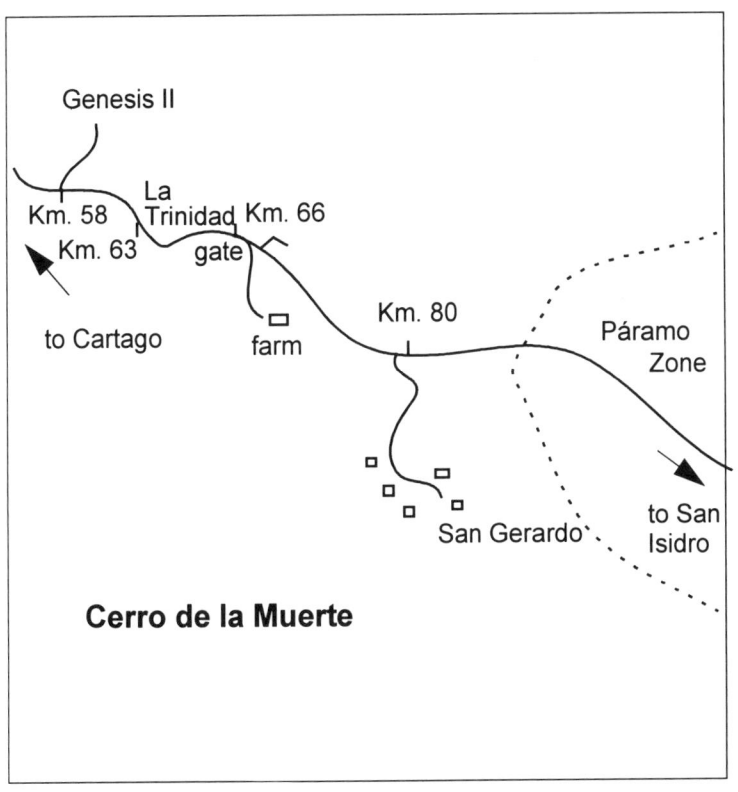

The first point of reference is the Abastador La Trinidad at Km. 63, where the key to the gate allowing you access to the nice oak forest at Km. 66 is kept. The entrance fee is 600 Colones per person, but is worth it for the Resplendent Quetzal and other species present. The road itself is on the right, and has a black gate with a small green plaque telling you to go to the *pulpería* if you want the key. Just up the highway, on the other side, a dirt road goes off toward some electric towers through shorter forest with good patches of brush and bamboo for Silvery-fronted Tapaculo, Black-billed Nightingale-Thrush, Wrenthrush, and other high-elevation specialties.

About Km. 70 a road to the right leads to a hotel and resturant with a Quetzal-locating guide service. The hotel is rather rustic.

At Km. 80 a gravel road goes down to the right 8 km to San Gerardo de Dota. There are several hotels at the end of the road. The most expensive hotel has trails into good Quetzal habitat, though you must climb quite far through the apple orchards and pasture to reach the forest. Otherwise there is some good birding along the road.

Around Km. 88 the páramo zone is reached, and the pass itself is shortly thereafter. Here, if conditions are good, you might find Timberline Wren, Wrenthrush, Volcano Hummingbird, and Volcano Junco.

Around Villa Mills at about Km. 95, the road starts down towards the Valle del General. The forest gets taller where it hasn't been converted into pasture. Anywhere you can find habitat is worth birding. There is a resturant and hotel right on the highway here.

At Km. 58 is the entrance to Genesis II, a private forest reserve. The house is about 4 km from the highway with every intersection well-marked. Admission is $7.50 to an good trail system through nice oak forest. Accommodations are available as well.

## Birds of Cerro de la Muerte

| | |
|---|---|
| Highland Tinamou | Swallow-tailed Kite |
| Red-tailed Hawk | Band-tailed Pigeon |
| Buff-fronted Quail-Dove | Sulphur-winged Parakeet |
| Barred Parakeet | White-crowned Parrot |
| Andean Pygmy-Owl | Bare-shanked Screech-Owl |
| Dusky Nightjar | Green Violetear |
| Fiery-throated Hummingbird | Variable Mountain-gem |
| Magnificent Hummingbird | Volcano Hummingbird |
| Collared Trogon | Resplendent Quetzal |
| Prong-billed Barbet | Emerald Toucanet |
| Acorn Woodpecker | Hairy Woodpecker |
| Ruddy Treerunner | Streak-breasted Treehunter |
| Spot-crowned Woodcreeper | Silvery-fronted Tapaculo |
| Golden-bellied Flycatcher | White-fronted Tyrannulet |
| Mountain Elaenia | Yellowish Flycatcher |
| Black-capped Flycatcher | Tufted Flycatcher |
| Dark Pewee | Ochraceous Pewee |

| | |
|---|---|
| Blue-and-White Swallow | Silvery-throated Jay |
| Ochraceous Wren | House Wren |
| Timberline Wren | Gray-breasted Wood-Wren |
| Black-billed Nightingale-Thrush | Ruddy-capped Nightingale-Thrush |
| Mountain Thrush | Sooty Thrush |
| Black-and-Yellow Silky-Flycatcher | Long-tailed Silky-Flycatcher |
| Yellow-winged Vireo | Brown-capped Vireo |
| Rufous-browed Peppershrike | Flame-throated Warbler |
| Black-throated Green Warbler | Wilson's Warbler |
| Slate-throated Whitestart | Collared Whitestart |
| Black-cheeked Warbler | Wrenthrush |
| Common Bush-Tanager | Sooty-capped Bush-Tanager |
| Yellow-thighed Finch | Large-footed Finch |
| Peg-billed Finch | Slaty Flower-piercer |
| Rufous-collared Sparrow | Eastern Meadowlark |

## San Vito area

Situated near the border, San Vito has several species more typical of Panama than Costa Rica, as well as good mid-elevation birding. Most of the area is cut over, but species variety is good in second-growth and at the Las Cruces Botanical Gardens.

San Vito is about 300 km from San José, or about five to six hours driving time over the Cerro de la Muerte, through San Isidro, and across the bridge at Paso Real to the town. Driving time depends on conditions on the Cerro.

There are a variety of hotels in San Vito, or you can stay at the OTS research facility at Las Cruces. A day visit to the gardens and adjacent forest reserve is $8 per person. You can spend the night for about $130 double in the high season, meals included. For reservations contact the Organization for Tropical Studies office in San José (see page 79).

Of considerable interest are some ponds near the San Vito airport, on the road to Sabalito. Arriving from Paso Real at the five-way intersection in downtown San Vito, go straight across the intersection, to the right of the La Ceiba hotel. After a few kilometers the airstrip will appear on the left. There is one pond about halfway down it, then a wet-

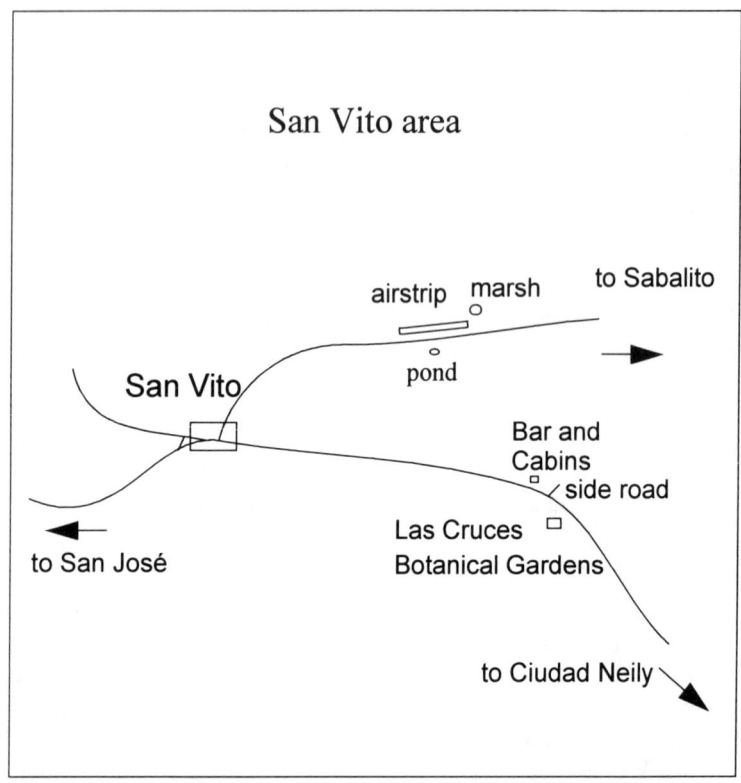

lands reserve at the far end. Walk along the path between the houses to get to the park. (There's even an observation tower.) Specialties include Masked Duck, Pale-breasted Spinetail, and Masked (Chiriquí) Yellowthroat.

Birding is also excellent around the Las Cruces botanical gardens, though you don't actually have to go in to see good birds. From the five-way intersection in San Vito, take the right fork about 5 km to the gardens. The well-tended gardens and the trails through the OTS reserve are well worth working. Also, a dirt road about 400 meters back towards San Vito and across the road has been productive in the past. This road leads about 300 meters to a small *campesino* house, whose occupants are unconcerned about birdwatchers. There is also a small bar and restaurant about a kilometer back towards San Vito.

## Birds of the San Vito area

Green Heron
Masked Duck
White-throated Crake
Purple Gallinule
Crimson-fronted Parakeet
White-crowned Parrot
Vaux's Swift
Rufous-tailed Hummingbird
Blue-crowned Motmot
Red-crowned Woodpecker
Buff-throated Foliage-gleaner
Buff-throated Woodcreeper
Russet Antshrike
Slaty Antwren
Lesser Elaenia
Ochre-bellied Flycatcher
Common Tody-Flycatcher
Golden-crowned Spadebill
Dusky-capped Flycatcher
White-ruffed Manakin
Plain Wren
Slaty-backed Nightingale-Thrush
White-throated Thrush
Tropical Parula
Kentucky Warbler
Wilson's Warbler
Bananaquit
Speckled Tanager
Golden-hooded Tanager
Blue Dacnis
Thick-billed Euphonia

Blue-winged Teal
Swallow-tailed Kite
Common Moorhen
Scaled Pigeon
Orange-chinned Parakeet
Striped Cuckoo
Little Hermit
Long-billed Starthroat
Violaceous Trogon
Pale-breasted Spinetail
Ruddy Foliage-gleaner
Streak-headed Woodcreeper
Plain Antvireo
Paltry Tyrannulet
Yellow-bellied Elaenia
Scale-crested Pygmy-Tyrant
Yellow-olive Flycatcher
Bright-rumped Attila
Piratic Flycatcher
Rufous-breasted Wren
Southern Nightingale Wren
Clay-colored Thrush
Golden-winged Warbler
American Redstart
Masked Yellowthroat
Rufous-capped Warbler
Silver-throated Tanager
Bay-headed Tanager
Scarlet-thighed Dacnis
Green Honeycreeper
Blue-gray Tanager

Red-crowned Ant-Tanager
Common Bush-Tanager
Buff-throated Saltator
Orange-billed Sparrow
Blue-black Grassquit
Yellow-faced Grassquit

Summer Tanager
Streaked Saltator
Black-headed Brush-Finch
Black-striped Sparrow
Variable Seedeater

## Monteverde Biological Reserve

Monteverde has established itself as one of the premier natural history destinations in Costa Rica. Overconstruction, a bad road, and often-inclement weather detract from the fabulous cloud forest reserve. There is a variety of accomodations available and the ambience is pleasant.

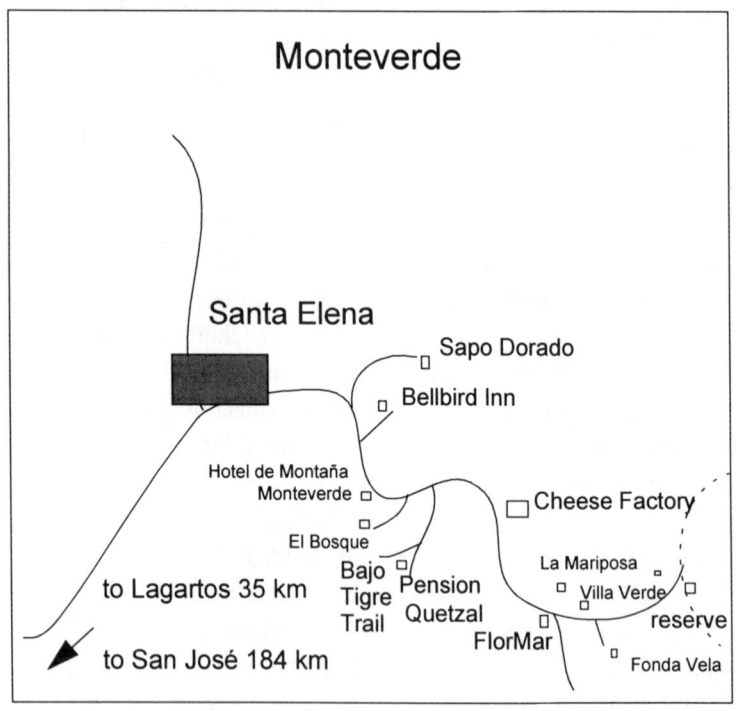

Santa Elena and the town of Monteverde are about 180 km from San José, or 3 hours. The last 35 km is over a very rough road. Keep an eye on your tires; if you get a flat you can get it fixed in Santa Elena. Just before Santa Elena there is a voluntary toll booth, where you can give them 100 Colones for the maintenance of the road if you feel like it. Don't feel obligated. There is a useful map where the road to Monteverde takes off.

There is now quite a variety of hotels between Santa Elena and the reserve, ranging in price from about $7 to $70. Construction has apparently outstripped demand, so you shouldn't have a problem without reservations, even in the high season.

The reserve itself, run by the Tropical Science Center, is one of the original efforts at conservation in the country. It is a magnificent piece of cloud forest with several discernible habitats. It costs $8 to get in. No more than 125 people are permitted in the reserve at any one time, so there is a very remote possibility you might have to wait. There are also hummingbird feeders by the gallery just down the road.

The reserve runs refuges for hikers in the Peñas Blancas valley. They are 3 and 6 hours hiking time from the headquarters and cost $3.50 per person. Boot rental is also available.

A good reference to the birds of the Monteverde area is the *Checklist of the Birds of Monteverde and Peñas Blancas* by Michael Fogden. It has abundance information for each of seven elevational zones on both slopes. The following list is primarily for the reserve itself, Zones 3 and 4 in the Fogden checklist.

## Birds of Monteverde

| | |
|---|---|
| Highland Tinamou | Black-chested Hawk |
| Swallow-tailed Kite | Barred Forest-Falcon |
| Black Guan | Black-breasted Wood-Quail |
| Band-tailed Pigeon | Ruddy Pigeon |
| Buff-fronted Quail-Dove | White-tipped Dove |
| Brown-hooded Parrot | Red-fronted Parrotlet-r |
| Squirrel Cuckoo | Bare-shanked Screech-Owl |
| Mottled Owl | Dusky Nightjar |

| | |
|---|---|
| Chestnut-collared Swift | Vaux's Swift |
| White-collared Swift | Coppery-headed Emerald |
| Green-fronted Lancebill | Fiery-throated Hummingbird |
| Green-crowned Brilliant | Green Hermit |
| Green Violetear | Magenta-throated Woodstar |
| Magnificent Hummingbird | Variable Mountain-gem |
| Scintillant Hummingbird | Stripe-tailed Hummingbird |
| Violet Sabrewing | Orange-bellied Trogon |
| Resplendent Quetzal | Prong-billed Barbet |
| Emerald Toucanet | Golden-olive Woodpecker |
| Hairy Woodpecker | Smoky-brown Woodpecker |
| Olivaceous Woodcreeper | Spotted Woodcreeper |
| Buffy Tuftedcheek | Lineated Foliage-gleaner |
| Red-faced Spinetail | Ruddy Treerunner |
| Spotted Barbtail | Streak-breasted Treehunter |
| Immaculate Antbird | Slaty Antwren |
| Silvery-fronted Tapaculo | Long-tailed Manakin |
| Bright-rumped Attila | Masked Tityra |
| Three-wattled Bellbird | Dark Pewee |
| Dusky-capped Flycatcher | Mountain Elaenia |
| Olive-striped Flycatcher | Paltry Tyrannulet |
| Sulphur-bellied Flycatcher | Tufted Flycatcher |
| Yellowish Flycatcher | White-throated Spadebill |
| Blue-and-White Swallow | Azure-hooded Jay |
| Gray-breasted Wood-Wren | Ochraceous Wren |
| House Wren | Black-faced Solitaire |
| Black-headed Nightingale-Thrush | Slaty-backed Nightingale-Thrush |
| Ruddy-capped Nightingale-Thrush | Swainson's Thrush |
| Clay-colored Thrush | Mountain Thrush |
| White-throated Thrush | Black-and-Yellow Silky-Flycatcher |
| Rufous-browed Peppershrike | Brown-capped Vireo |
| Lesser Greenlet | Slaty Flower-piercer |
| Scarlet-thighed Dacnis | Black-and-White Warbler |

Black-throated Green Warbler
Golden-winged Warbler
Wilson's Warbler
Slate-throated Whitestart
Three-striped Warbler
Northern Oriole
Common Bush-Tanager
Hepatic Tanager
Spangle-cheeked Tanager
Buff-throated Saltator
Yellow-throated Brush-Finch
Yellow-thighed Finch
Rufous-collared Sparrow

Townsend's Warbler
Blackburnian Warbler
Collared Whitestart
Golden-crowned Warbler
Wrenthrush
Golden-browed Chlorophonia
Sooty-capped Bush-Tanager
Silver-throated Tanager
Black-thighed Grosbeak
Chestnut-capped Brush-Finch
Sooty-faced Finch
White-eared Ground-Sparrow
Yellow-faced Grassquit

# The Atlantic Lowlands

# The Atlantic Lowlands

The Atlantic slope of Costa Rica, until recently mostly forest, has been converted to cattle pasture and banana plantations with disconcerting speed. Despite the large portion of the country it represents, there are few good, easily-accessible birding spots. Some of the interesting species can also be seen at Caribbean foothill localities.

Most of the avifauna of the Caribbean lowlands is widespread, so if you have spent time in eastern Mexico or lowland South America this area might be of low priority. Nonetheless, it is essential to visit if you want a big trip list, and the beaches are very nice.

## Finca La Selva

Owned by the Organization for Tropical Studies, a consortium of universities dedicated to research on the workings of the tropical rain forest, La Selva offers much for the birder. It is not a tourist facility. Prices are high and access is limited, but it's worth the trouble.

La Selva is near the town of Puerto Viejo de Sarapiquí in Heredia province. (Don't confuse this with Puerto Viejo on the coast.) Access is via the Braulio Carrillo highway and a paved road to Horquetas and Puerto Viejo. Watch for signs that say "OET," the Spanish initials. Several nearby lodges are a possible source of more-affordable accommodations, but none are affiliated with the reserve.

Accommodation at the research station is mostly in dormitory rooms, and meals are in the cafeteria. Prices for a double in 1996 were $120 per night, with three meals included. A more economical solution, if you have transport, is to stay in Puerto Viejo and pay $20 per person for a 3½ hour guided tour, offered at 8 a.m. and 1:30 p.m. Note that these tours are for tourists, and are not likely to be conducive to birding; inquire if they have added something for birdwatchers.

In either case you must make advance reservations with the Organization for Tropical Studies, and space is limited. If you stay at the station, be sure that your reservation is for dinner, breakfast, and lunch the next day, to guarantee access to the forest in the morning; otherwise an arriving group might have priority over you. Schedule this visit first and design the rest of your trip around it.

(from outside Costa Rica)
Organization For Tropical Studies
Interlink 341
Box 025635
Miami, FL 33152
(506) 240-6696, fax 240-6783, e-mail reservas@ns.ots.ac.cr
Make reservations for day tours directly with the field station at 710-1515, fax 710-1414.

The reserve has an excellent trail system and a huge birdlist. Upon check-in, you will get a map showing trails and areas with active research projects to avoid. It is essential to spend as much time as possible in the forest. Tape recorders will most likely be banned soon, and birding can be difficult. Persistence will pay off here perhaps more than any other location in Costa Rica.

## Birds of La Selva

| | |
|---|---|
| Great Tinamou | Little Tinamou |
| Slaty-breasted Tinamou | Broad-winged Hawk |
| Semiplumbeous Hawk | Black Hawk-Eagle |
| Crested Guan | White-throated Crake |
| Sungrebe | Short-billed Pigeon |
| Gray-chested Dove | Olive-backed Quail-Dove |
| Olive-throated Parakeet | Crimson-fronted Parakeet |
| White-crowned Parrot | Brown-hooded Parrot |
| Red-lored Parrot | Mealy Parrot |
| Squirrel Cuckoo | Vermiculated Screech-Owl |
| Spectacled Owl | Least (C. A.) Pygmy-Owl |
| Pauraque | White-collared Swift |
| Gray-rumped Swift | L. Swallow-tailed Swift |
| Bronzy Hermit | Long-tailed Hermit |
| Little Hermit | White-necked Jacobin |
| Violet-crowned Woodnymph | Blue-chested Hummingbird |
| Snowcap | Bronze-tailed Plumeleteer |
| Slaty-tailed Trogon | Black-throated Trogon |

Violaceous Trogon
Broad-billed Motmot
Rufous-tailed Jacamar
White-whiskered Puffbird
Collared Aracari
Chestnut-mandibled Toucan
Chestnut-colored Woodpecker
Black-cheeked Woodpecker
Plain-brown Woodcreeper
Barred Woodcreeper
Black-striped Woodcreeper
Slaty Spinetail
Great Antshrike
Streak-crowned Antvireo
Checker-throated Antwren
Chestnut-backed Antbird
Fulvous-bellied Antpitta
Snowy Cotinga
Rufous Piha
Cinnamon Becard
Masked Tityra
Red-capped Manakin
Long-tailed Tyrant
Tropical Pewee
Ruddy-tailed Flycatcher
Black-headed Tody-Flycatcher
Slate-headed Tody-Flycatcher
Northern Bentbill
Paltry Tyrannulet
Ochre-bellied Flycatcher
Bay Wren
White-breasted Wood-Wren
Tropical Gnatcatcher

Green Kingfisher
Rufous Motmot
Pied Puffbird
White-fronted Nunbird
Keel-billed Toucan
Rufous-winged Woodpecker
Cinnamon Woodpecker
Pale-billed Woodpecker
Wedge-billed Woodcreeper
Buff-throated Woodcreeper
Streak-headed Woodcreeper
Buff-throated Foliage-Gleaner
Slaty Antshrike
White-flanked Antwren
Dusky Antbird
Black-faced Antthrush
Spectacled Antpitta
Bright-rumped Attila
Rufous Mourner
White-winged Becard
Purple-throated Fruitcrow
White-collared Manakin
White-ringed Flycatcher
Yellow-bellied Flycatcher
Golden-crowned Spadebill
Common Tody-Flycatcher
Yellow-margined Flycatcher
Black-capped Pygmy-Tyrant
Brown-capped Tyrannulet
Band-backed Wren
Stripe-breasted Wren
Song Wren
Long-billed Gnatwren

| | |
|---|---|
| Tawny-faced Gnatwren | Green Shrike-Vireo |
| Lesser Greenlet | Tawny-crowned Greenlet |
| Shining Honeycreeper | Bananaquit |
| Blackburnian Warbler | Chestnut-sided Warbler |
| Northern Waterthrush | Ovenbird |
| Kentucky Warbler | Olive-crowned Yellowthroat |
| Gray-crowned Yellowthroat | Buff-rumped Warbler |
| Montezuma Oropendola | Scarlet-rumped Cacique |
| Yellow-billed Cacique | Yellow-crowned Euphonia |
| Olive-backed Euphonia | Plain-colored Tanager |
| Golden-hooded Tanager | Crimson-collared Tanager |
| Summer Tanager | Olive Tanager |
| Tawny-crested Tanager | Red-throated Ant-Tanager |
| Dusky-faced Tanager | Black-headed Saltator |
| Buff-throated Saltator | Black-faced Grosbeak |
| Blue-black Grosbeak | Yellow-faced Grassquit |
| Variable Seedeater | Lesser Seed-Finch |
| Black-striped Sparrow | Orange-billed Sparrow |
| Slate-colored Grosbeak | |

## Cahuita and Puerto Viejo

The Caribbean coast south of Limón provides an easily accessible introduction to the birds of the Atlantic slope, without the expense or hassle of some of the other areas. The well-developed tourist areas of Cahuita and Puerto Viejo are the obvious centers for exploration. Though some distance apart, they are combined here as the birds are similar. Here too you can enjoy some fine beaches.

The high season of 1994-95 saw a number of crimes against tourists, including two murders. This situation was much exaggerated by the San José press and the tourist industry in other parts of the country. Nonetheless, as similar incidents have happened in the past, it is worth taking certain precautions in the area. I cannot recommend you drive the highway from Limón to Cahuita at night, or arrive in either town after dark. The driving times below should be adequate to get you there; keep them in mind before you leave San José. If necessary you can stay in Limón, but don't wander around there at night either.

Cahuita and Puerto Viejo are readily reached by the coastal strand highway that begins in Limón. It is the first main road on the right, just before the docks as you enter town from San José. Maybe the sign for Cahuita and the Limón airport will be replaced someday. You will know you're on the right road when you cross a grotty looking estuary into the similarly grotty town of Cieneguita. Via the Braulio Carrillo highway Limón is about a 2½ hour drive from San José.

Along the 43 km drive from Limón to Cahuita there are several river mouths that have a few shorebirds, terns, etc. in season. The town itself is about a kilometer off the highway towards the ocean, so watch for signs. There is a variety of accommodation available in Cahuita.

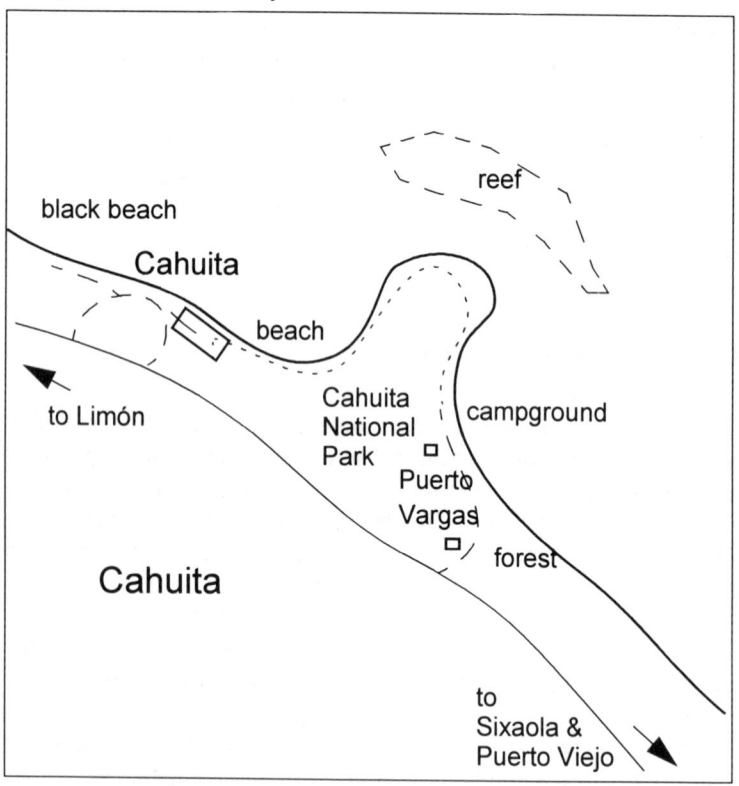

Camping is available on the Puerto Vargas side of the park, but due to problems of theft and assault, it is not recommended. Even if you are on an extremely low budget, prices in town are affordable.

After locals (dependent on the budget travelers that flock to Cahuita) staged a mini-rebellion to protest the park fee increase, the park service quit charging to walk from town into the park along the beach. You still have to pay to enter from Puerto Vargas.

Much of the best birding in the area is from the highway, with open-country species and some local specialties such as Nicaraguan Seed-Finch and Blue-headed Parrot. Cahuita National Park is primarily a marine park, and doesn't have much in the way of trails except for the beach from Cahuita town to Puerto Vargas. There is one short trail inland starting just inside the park boundary on the Cahuita side. The road to Puerto Vargas and the campground goes through nice forest.

Sixteen km further down the road is Puerto Viejo, popular with foreign tourists and surfers. Hotels of all shapes are springing up everywhere in town and along the road to Manzanillo. Prices are mostly modest. Birding here in generally similar to Cahuita, especially along the road to Manzanillo or Punta Mona. Of interest here are the abandoned cacao plantations in the hills above town, most of which are being filled with houses as the growth of tourism brings more people. A good trail leaves from the southwest corner of the soccer field on the back side of town. Other trails might be located with local inquiry.

Along the road to Manzanillo a side road goes back into the hills with some good forest among the fields. It is about 6.5 km from Puerto Viejo on the right, with a sign that says "Punta Uva Caminata guiada al bosque." Don't try to drive across the bridges, because they were damaged in the earthquake that hit Limón province in 1991. There is horse rental available if you wish to cover more ground.

## Birds of Cahuita and Puerto Viejo

| | |
|---|---|
| Little Tinamou | Common Black-Hawk |
| Broad-winged Hawk | Gray-headed Chachalaca |
| White-throated Crake | Gray-necked Wood-Rail |
| Northern Jacana | Royal Tern |
| Pale-vented Pigeon | Short-billed Pigeon |

| | |
|---|---|
| Blue Ground-Dove | White-tipped Dove |
| Gray-chested Dove | Olive-backed Quail-Dove |
| Crimson-fronted Parakeet | Blue-headed Parrot |
| Mealy Parrot | Squirrel Cuckoo |
| Gray-rumped Swift | Bronzy Hermit |
| Long-tailed Hermit | Little Hermit |
| Blue-chested Hummingbird | Rufous-tailed Hummingbird |
| Bronze-tailed Plumeleteer | Violaceous Trogon |
| Black-throated Trogon | Ringed Kingfisher |
| Belted Kingfisher | Green Kingfisher |
| Collared Aracari | Keel-billed Toucan |
| Black-cheeked Woodpecker | Lineated Woodpecker |
| Plain-brown Woodcreeper | Barred Woodcreeper |
| Buff-throated Woodcreeper | Streak-headed Woodcreeper |
| Fasciated Antshrike | Slaty Antshrike |
| Checker-throated Antwren | Dot-winged Antwren |
| Dusky Antbird | Chestnut-backed Antbird |
| Ochre-bellied Flycatcher | Common Tody-Flycatcher |
| Black-headed Tody-Flycatcher | Tropical Pewee |
| Long-tailed Tyrant | Bright-rumped Attila |
| Dusky-capped Flycatcher | Masked Tityra |
| Snowy Cotinga | White-collared Manakin |
| Gray-breasted Martin | S. Rough-winged Swallow |
| Band-backed Wren | Black-throated Wren |
| Plain Wren | Long-billed Gnatwren |
| Tropical Gnatcatcher | Clay-colored Thrush |
| Yellow-throated Vireo | Lesser Greenlet |
| Tennessee Warbler | Chestnut-sided Warbler |
| Bay-breasted Warbler | Prothonotary Warbler |
| Northern Waterthrush | Kentucky Warbler |
| Olive-crowned Yellowthroat | Bananaquit |
| Golden-hooded Tanager | Red-legged Honeycreeper |
| Yellow-crowned Euphonia | Olive-backed Euphonia |

| | |
|---|---|
| White-shouldered Tanager | Tawny-crested Tanager |
| Sulphur-rumped Tanager | White-lined Tanager |
| Red-throated Ant-Tanager | Summer Tanager |
| Crimson-collared Tanager | Scarlet-rumped Tanager |
| Black-headed Saltator | Rose-breasted Grosbeak |
| Blue-black Grosbeak | Black-striped Sparrow |
| Blue-black Grassquit | Variable Seedeater |
| White-collared Seedeater | Lesser Seed-Finch |
| Giant Cowbird | Black-cowled Oriole |
| Orchard Oriole | Northern Oriole |
| Montezuma Oropendola | Chestnut-headed Oropendola |

## Tortuguero National Park

This national park is popular with nature tourists for its nesting sea turtles and good birding. Unfortunately access is still limited following damage to the inland canal in the 1991 earthquake. Birding and the jungle experience might make a tour worthwhile, though there are few species that cannot be seen more easily elsewhere in Costa Rica. The boat trip might produce Sungrebe, Green-and-Rufous Kingfisher, and Rufescent Tiger-Heron.

Tours run about $150 for three days/two nights, though with a little shopping around you should find something for less. See any travel agent in San José for details and current schedules.

## Caño Negro Biological Reserve

This important wetland is known mainly for its concentrations of waterbirds at the end of the dry season as the lagoon slowly dries out. Many of the birds that inhabit the Guanacaste marshes move here as those areas dry out. The waterbird species are the same as can be found almost anywhere else in the American tropics or subtropics, though Caño Negro does have a population of the local Nicaraguan Grackle.

To see the larger numbers of birds you really need to take a boat, and there have been regular tours leaving from Los Chiles, or arranged by agencies in Fortuna or San José. Schedules and prices are difficult to predict, since there has been a fall-off in demand following incidents of kidnapping for ransom in the Northern Zone.

The combination of low enough water levels to attract the birds and high enough for the boats to enter the lagoon is somewhat limited, mostly to December and January. After that, it is possible to see much of the lagoon (with a scope) from the village of Caño Negro on the north side of the lake. After May or so, flooding makes the area unattractive.

The boat tours mostly start along the Río Frio near Los Chiles, easily reached by paved highway from San José. To get to the town of Caño Negro, however, you must come in on a gravel road from the Upala side. The road that some maps show as connecting Caño Negro and Los Chiles has no bridge over the Río Frio.

## Birds of Caño Negro

| | |
|---|---|
| Neotropic Cormorant | Anhinga |
| Great Blue Heron | Great Egret |
| Snowy Egret | Little Blue Heron |
| Tricolored Heron | Green Heron |
| White Ibis | Roseate Spoonbill |
| Fulvous Whistling-Duck | Black-bellied Whistling-Duck |
| Muscovy | Blue-winged Teal |
| Lesser Yellow-headed Vulture | Osprey |
| Black-collared Hawk | White-throated Crake |
| Gray-necked Wood-Rail | Purple Gallinule |
| Sungrebe | Solitary Sandpiper |
| Spotted Sandpiper | Lesser Yellowlegs |
| Least Sandpiper | Pale-vented Pigeon |
| White-tipped Dove | Olive-throated Parakeet |
| White-crowned Parrot | Lesser Nighthawk |
| Rufous-tailed Hummingbird | Black-headed Trogon |
| Ringed Kingfisher | Green Kingfisher |
| Amazon Kingfisher | Slaty Spinetail |
| Barred Antshrike | Yellow-bellied Tyrannulet |

Yellow-bellied Elaenia
Tropical Pewee
Gray-breasted Martin
Barn Swallow
Prothonotary Warbler
Bananaquit
Buff-throated Saltator
Lesser Seed-Finch
Nicaraguan Grackle
Montezuma Oropendola

Common Tody-Flycatcher
White-collared Manakin
Mangrove Swallow
Yellow Warbler
Northern Waterthrush
Yellow-throated Euphonia
Black-striped Sparrow
Red-winged Blackbird
Yellow-billed Cacique

# Panama

Panama is a country with a great variety of habitats and birdlife within its borders; it deserves more attention from the foreign birder than it has received in recent years. The historical accident of the canal with its attendant militarized zone has resulted in good forested habitats close to one of the most cosmopolitan cities in Latin America.

Unfortunately, from the perspective of the conservationist, the view is not rosy. The former Canal Zone is gradually being turned over to the Panamanian government, and powerful economic interests are eyeing what can only be considered some of the most valuable undeveloped real estate in the world. Even areas currently designated as national parks are not considered safe by the local conservation movement. So bird Achiote Road, Pipeline Road, and other areas near Panama City and Colon now while you have the chance.

## When to Come

The dry season from December to April has the nicest weather, but birding is good throughout the year. The dry season also coincides with the largest numbers of North American migrants, to help pad the species total. Note that Panama is rather far west in its time zone, so dry season birding often does not get underway until after 6 a.m.

In the lowlands, most rainy season precipitation comes in the form of thundershowers in the afternoon, when the birding isn't as good anyway. Mud can be a factor. Away from the lowlands, weather is more likely to harm your birding. The Volcán Barú area tends to have good weather, but other highland areas can be affected at any time. March seems to be the best time to visit areas like eastern Chiriquí, and on the balance is probably the best month to visit if you can avoid Easter.

## International Transport

The Omar Torrijos International Airport outside of Panama City is one of the better airports in Latin America in terms of its range of destinations available. (One would perhaps wonder why they named the airport after someone who was killed in a plane crash.) Panama is well served from Miami via American, Lloyd Air Boliviano, and Copa, with Houston on Continental also a possibility. Copa and others also serve a variety of cities in South America and the Caribbean.

Citizens of the U.S. need either a free visa or a tourist card ($5) to stay up to 30 days. Requirements for other nationalities vary, so check.

## Domestic Flights

There are also several airlines serving destinations in the "interior" of Panama. Except for perhaps a trip to the Darién or David, these are not of much interest to the birder. Note that domestic flights leave from a different place: Paitilla airport near the city.

## Busses

Public transport reaches even the smallest towns in the provinces, and is extensive in the cities. Except for a few express routes, most service is with smaller busses than those used elsewhere. Service is frequent and cheap, but reaches almost none of the actual birding areas. To bird Panama, you are definitely better off with a car.

## Rental Cars

Car rental is relatively cheap in Panama compared with most of Latin America. The major chains are well-represented at the airports and elsewhere in Panama City, as well as in David and some of the other larger towns. Prices are lower if reservations are made from abroad. For reference, prices from Dollar in June of 1996 were quoted at $131 per week for a small car.

Gasoline is about $1.65-$1.75 per gallon (and sold that way). Fuel is widely available.

## Driving Standards

With the help of good roads, driving standards are not too bad in Panama, perhaps comparable to New Jersey. There are few of the treacherous mountain roads that make driving such an adventure in neighboring countries.

Panama City is an easy place to get lost, but that is due more to a lack of signs on major thruways than anything else. The smaller streets usually have signs.

The Pan-American highway west towards Costa Rica is generally in good shape, with occasional rough spots. Some parts pass through sugar cane growing areas with slow truck and tractor traffic, but for the most part one can average close to 100 kph outside the towns.

Speed limits often are set ridiculously low, and are generally disregarded by the driving public. Most of the highways have 80 or 90 kph speed limits, but it is virtually never marked where a town's speed zone ends and the open highway begins. On the highway the transit police wear orange vests and stand near their vehicles, so they are easy to see.

Traffic laws are mainly enforced when the police want to shake you down for a bribe. They cannot by law make you pay on the spot; the best thing to do is play dumb and wait them out. Not that you would know the delicate bribery etiquette anyway. Even if a ticket has been written, it often will not be given to you.

## Accommodations

Hotels in Panama are abundant, and usually good value. Away from the central cities there tends to be fewer choices, but $20-25 will usually get you an air-conditioned room with hot water, and a TV and phone that you don't want. Be sure to ask for the quietest room available, as many hotels are near the highway. Beware any place where walls block the view of the parking from the road, as these are short-time places. Advertisements of 24-hour service are not a good sign.

## Food

Panama's national cuisine is of little interest to the international traveler. Restaurant food is usually tasty and good value if a little greasy. Many highway restaurants open early or even stay open 24 hours, so there is often a chance for an early breakfast. There are several 24 hour restaurants in downtown Panama City.

## Money

The currency of Panama is the Balboa, interchangeable with the U.S. Dollar. In fact, all the paper money is American currency. Sometime you will encounter someone who doesn't want to call them *dólares*, but I have examined the notes in circulation closely, and none bears the word Balboa anywhere. U.S. coins circulate alongside Panamanian ones, which are of the same denominations and sizes.

This is of course a great convenience for the traveler, as you don't have to worry about changing money. Large bills like fifties or hundreds are treated with suspicion, due to the amount of counterfeit money in circulation. Visa and Mastercard are widely accepted, other cards less so. Travelers' checks are also a good bet.

## Safety

Panama City and especially Colon are potentially dangerous cities, but with common sense you can avoid any trouble. The countryside is almost invariably peaceful.

Colon is one of the worst cities in the Americas for crime, and failure to heed warnings will likely be regretted. Unfortunately, you will probably need to stay there to bird the Caribbean side of the canal area. Lock your doors as you drive through town, and under no circumstances walk around even in the daytime.

For the most part your car and contents will be safe when birding, though be sure not to leave anything in view from outside to tempt an opportunistic thief. Some areas of the Chiriquí highlands have had a problem with car break-ins in the past.

See the Costa Rica section for a discussion of safety in the field.

## The Panamanians

The people of central Panama are such a diverse lot that generalizations are difficult. Panama City's position as a trade and banking center has resulted in a cosmopolitan populace.

There is some resentment for the treatment Panama has received at the hands of the U.S. government and military, but this is not always evident against the background of big-city unfriendliness in Panama City. In the countryside, people are likely to be helpful or even curious. Several birding spots are within Indian reservations, and you can expect an extra dose of staring there.

## Language

Many people in the Canal area speak English, some as their native language. In the countryside, only Spanish is the norm. While it is certainly helpful to speak Spanish, a traveler can get by without it.

## Health

Serious health risks are few in Panama, particularly in the areas mentioned in this book. Malaria is not a factor so don't take the pills. You'll almost certainly get some sort of intestinal disorder shortly after arrival, but for the most part food and water are safe. See the Costa Rica section for advice on this subject.

## Resources

The main book for birders in Panama is of course Robert Ridgely's *A Guide to the Birds of Panama*. Other books you might find useful include *The Birds of Colombia* by Steve Hilty *et. al.* and a *Field Guide to the Birds of Costa Rica* by Gary Stiles and Alexander Skutch.

Tourist guidebooks are mostly not too good. The *Central American Handbook* and Lonely Planet's *Central America on a Shoestring* have some coverage, but their treatment of hotels in Panama City is dismal.

It is also difficult to get good maps, so the one the rental car company gives you will have to do. Hopefully, the maps in this book will send you in the right direction, but don't be afraid to ask for directions.

The Panama Audubon Society has some checklists and such, and it is a good idea to join anyway, since it is one of the premier conservation organizations in Panama. Annual membership is $30 for residents outside Panama.

## Panama City

Getting around Panama City is a bit of an adventure for the uninitiated, though traffic flows fairly well. The main problem is that the major arterials are poorly marked; sometimes when lost it's easier to get off on the side streets where there are usually signs.

The birding locations are on every side of the city, so there's no really convenient area for accommodations. There is one hotel near the airport and Tocumen, but most of the reasonable accommodation is in the area of Av. Central/Via España/Av. Perú and Calles 25-50 or so. One disadvantage of this downtown area is that many of the cheaper places do a brisk business in short-term stays, which can result in more noise than the birder would perhaps like. This occurs in fairly nice places in addition to the obviously sleazier *pensiones*. So if you want to pay less then $30 or so, turn up the air-conditioner to drown out the racket! It will be hard to find a place with secure parking for less than that anyway.

Despite the drawbacks noted above, the Via España area is probably where you should stay. It also has a reasonable 24 hour restaurant, the Benidorm, on Calle 30 between Avenidas 1 & 2.

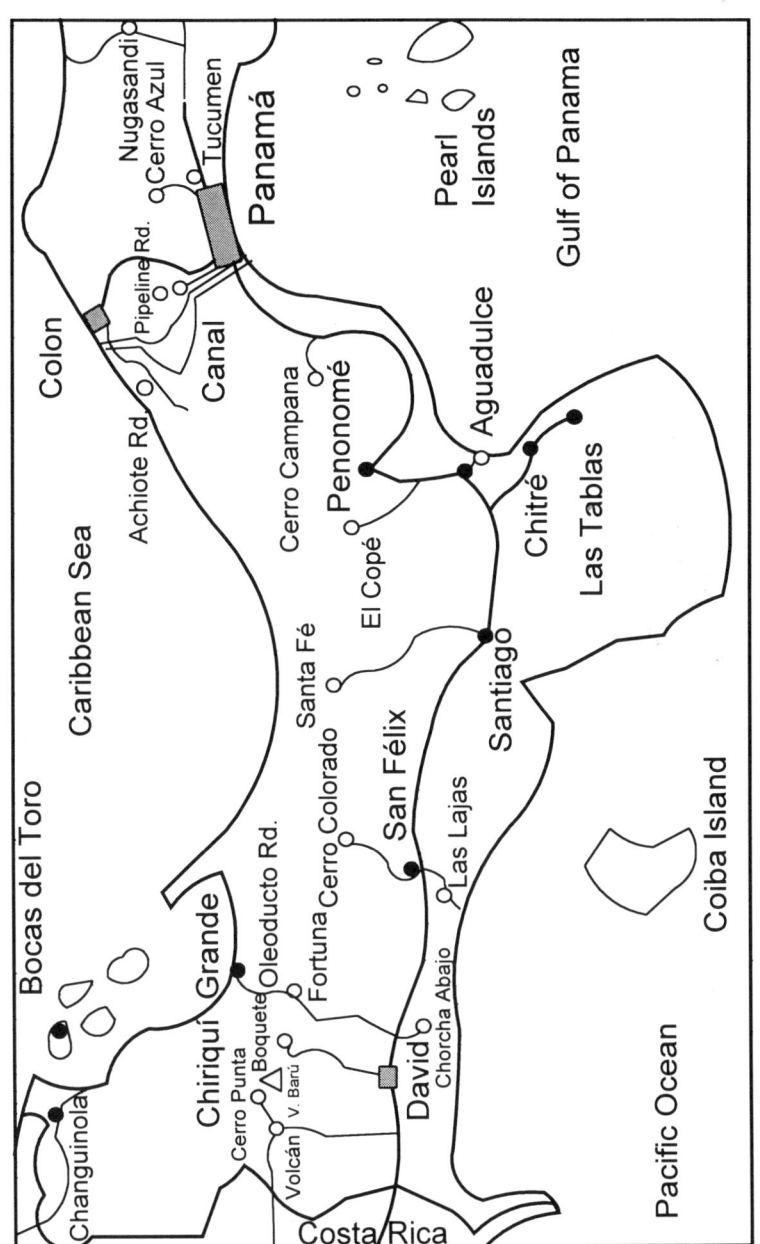

# Central Panama

Caribbean Sea
Central Panama
Galeta Island
Ft. San Lorenzo
Colon
Gatun
Madden Lake
Achiote Rd.
Pipeline Rd.
Cerro Azul
Gatun Lake
Madden Forest
Gamboa
Plantation Rd.
Tocumen
Chiva
Chiva Rd.
Summit Gardens
Panama City
Canal
to David
Pacific Ocean

to Fort Clayton and Gamboa
Panama City
to Colon
Av. Alfaro
Ancon hill
Av. Simón Bolívar
Av. 1 N
(Hotels)
Av. Bausta
Via Brasil
Via España
to Tocumen
Av. Balboa
Bay of Panama
Paitilla airport
Panama Viejo

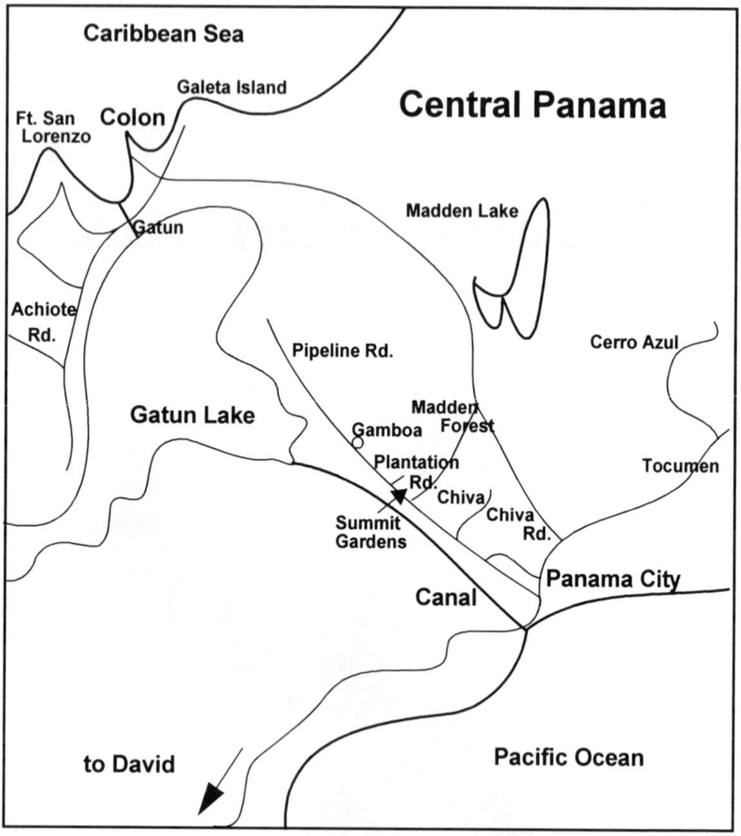

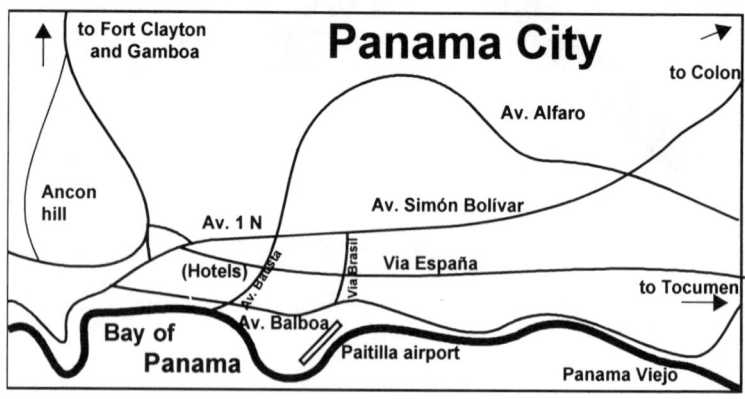

# Central Panama

The lowlands of Central Panama provide some of the most-accessible tropical forest in the world. Roads are good, accommodations are ample (though none terribly convenient), and a wide variety of birds can be seen in a small area.

As the old Canal Zone is gradually turned over to the Panamanian government, many of the areas in this section are losing the protection that the security needs of the canal provided. Much is now set aside as parks or canal catchment zones. Powerful economic interests are eyeing this land, and as the recent episode involving a road through the Metropolitan Nature Park demonstrated, legal protected status is not enough in the Panamanian political climate. Worse, the Pipeline Road forests lie on the logical route for a new highway from Panama to Colon. So plan to bird these areas while you have the chance.

## Panama City

Aside from its obvious standing as your base for birding Central Panama, the capital has some spots within city limits. It remains to be seen how the new road through the Metropolitan Nature Park will affect birding, though the park will likely remain a possibility for a brief visit.

The other main attraction in town is the shorebird flats of the Bay of Panama. Large numbers of shorebirds are present more or less year-round, with the highest numbers during August-September and March-April. Tidal variation is high in the bay, so the best time for birding is generally two hours on either side of high tide. Look in the local newspapers for times.

The two main areas for viewing are Panama Viejo and Juan Diaz. The ruins of Panama Viejo are right in the city, and can be reached from the Via Cincuentenario (see map).

The mouth of the Juan Diaz river and the nearby bay is also a possibility. Continue out Via España towards Tocumen until you see a large horse-racing track on the right. Take the next right on 116th street, and go straight through the residential area, then onto a gravel road to the beach. This road also passes through some interesting mangroves that have Mangrove Black-Hawk, American Pygmy Kingfisher, Straight-billed Woodcreeper, "Mangrove" Warbler, and Spot-breasted Woodpecker (rare).

# The Pacific side of the Canal area

Most Pacific-side birding spots are accessed from the Gaillard Highway, which leaves the west side of the city and goes north past Fort Clayton along the canal to Gamboa. Coming off Av. Central, take a right at the first major intersection. This intersection is totally unmarked, and you may have to scout it out in the daytime first. Use Ancon Hill for bearings; head towards it from the city and then go north.

The Madden Forest and Madden Dam are more easily reached from the Transisthmian Highway to Colon. Any of these areas are relatively close to Panama City, so they are good to fill holes in your itinerary.

## Chiva Chiva Road

About 2 km beyond Fort Clayton, the Chiva Chiva Road takes off to the right between two large ponds. There is a park sign and a usually unmanned entrance station there as well, and some traffic to a garbage dump. The ponds themselves are of interest, and the woodlands can be good. A few kilometers in, just beyond the power lines, a road goes off to the left into more good woodland.

## Summit Gardens

The botanical gardens here are good for easy birding if there aren't too many people. The picnic areas have hummingbirds and other open-area species, and there is a zoo with parrots and such.

## Plantation Road

The old Plantation road is now maintained as a trail, marked with a sign that says "Plantation Loop." There is access to good forest here, or along the paved road that leads up to a radar station on the top of the hill.

## Gamboa

There isn't too much birding around Gamboa but you will have to come out here to get your Pipeline Road permit, so you might look around at the marina.

## Madden Forest

This forest area is easily reached from either Gaillard Highway or the Transisthmian. From the Gaillard, stay straight at the big Parque Nacional Soberanía sign mentioned above and continue to the forest. From the Transisthmian Highway, watch for the overpass (ca. 20 km from Panamá) that crosses Madden Road, the only bridge on the highway over another road. The access to Madden Road is on the left, just to the Colon side of the bridge. Go right for Madden Forest and left for

Madden Dam.

Madden Road has many small side roads that give access to the forested areas; most are gated so you'll have to leave your car by the road.

**Madden Dam**

The primary birding area here is a gated road just to the Panamá side of the dam that creates Madden Lake. From the Madden Road intersection mentioned above, continue until the dam comes into sight. You have probably just passed the road on the right. From Colon, there is a fork, fairly well-marked for Madden Dam, by a police station. Cross the dam to the road, which continues to an old Scout Camp.

The first part of this road is mostly second-growth, but is good for Rosy Thrush-Tanager and others. There is better forest beyond, as well as views of the lake.

## Birds of the Pacific side

| | |
|---|---|
| Little Tinamou | Blue-winged Teal |
| White-tailed Kite | Mangrove Black-Hawk |
| Savannah Hawk | Roadside Hawk |
| Broad-winged Hawk | Crested Caracara |
| Yellow-headed Caracara | Laughing Falcon |
| Wattled Jacana | Pale-vented Pigeon |
| Scaled Pigeon | White-tipped Dove |
| Orange-chinned Parakeet | Red-lored Parrot |
| Yellow-crowned Parrot | Squirrel Cuckoo |
| Lesser Nighthawk | Short-tailed Swift |
| Lesser Swallow-tailed Swift | Rufous-breasted Hermit |
| Long-tailed Hermit | White-necked Jacobin |
| Violet-crowned Woodnymph | Violet-bellied Hummingbird |
| Snowy-bellied Hummingbird | Rufous-tailed Hummingbird |
| White-vented Plumeleteer | Violaceous Trogon |
| Black-throated Trogon | Slaty-tailed Trogon |
| Blue-crowned Motmot | Rufous Motmot |
| Broad-billed Motmot | Ringed Kingfisher |

White-necked Puffbird
Keel-billed Toucan
Red-crowned Woodpecker
Lineated Woodpecker
Buff-throated Woodcreeper
Barred Antshrike
Checker-throated Antwren
Dot-winged Antwren
White-bellied Antbird
Spotted Antbird
Paltry Tyrannulet
S. Beardless Tyrannulet
Yellow-crowned Tyrannulet
Yellow Tyrannulet
Common Tody-Flycatcher
Yellow-margined Flycatcher
Tropical Pewee
Bright-rumped Attila
Panama Flycatcher
Boat-billed Flycatcher
Social Flycatcher
Piratic Flycatcher
Masked Tityra
Golden-collared Manakin
Blue-crowned Manakin
Gray-breasted Martin
Black-chested Jay
White-breasted Wood-Wren
Long-billed Gnatwren
Clay-colored Thrush
Yellow-green Vireo
Golden-fronted Greenlet

Collared Aracari
Chestnut-mandibled Toucan
Black-cheeked Woodpecker
Plain Xenops
Fasciated Antshrike
Slaty Antshrike
White-flanked Antwren
Dusky Antbird
Chestnut-backed Antbird
Black-faced Antthrush
Brown-capped Tyrannulet
Mouse-colored Tyrannulet-r
Ochre-bellied Flycatcher
Pale-eyed Pygmy-Tyrant-r
Yellow-olive Flycatcher
Ruddy-tailed Flycatcher
Acadian Flycatcher
Dusky-capped Flycatcher
Great Kiskadee
Rusty-margined Flycatcher
Streaked Flycatcher
Fork-tailed Flycatcher
Black-crowned Tityra
Lance-tailed Manakin
Red-capped Manakin
Barn Swallow
Buff-breasted Wren
Song Wren
Tropical Gnatcatcher
Yellow-throated Vireo
Scrub Greenlet
Lesser Greenlet

| | |
|---|---|
| Tennessee Warbler | Yellow Warbler |
| Chestnut-sided Warbler | Bay-breasted Warbler |
| Prothonotary Warbler | Plain-colored Tanager |
| Golden-masked Tanager | Blue Dacnis |
| Green Honeycreeper | Red-legged Honeycreeper |
| Yellow-crowned Euphonia | Thick-billed Euphonia |
| White-shouldered Tanager | Crimson-backed Tanager |
| Rosy Thrush-Tanager | Streaked Saltator |
| Buff-throated Saltator | Black-striped Sparrow |
| Orchard Oriole | Northern Oriole |
| Yellow-backed Oriole | Crested Oropendola |
| Chestnut-headed Oropendola | Lesser Goldfinch |

## Pipeline Road

The forests of Pipeline Road are a tremendous location for a variety of birdlife, though they can be quiet too. Several different habitats are available along the nearly 20 km of road, so it takes more than one day to cover the area.

Access to Pipeline is restricted; you need to get written permission from the INRENARE office in Gamboa. The office can be located by taking the first left after you cross the one-way bridge at the edge of Gamboa, then bearing left at a Y, then taking a right into a residential neighborhood just after a playground. It is possible to walk into the park, but the early stretch of road is mostly second-growth; the better birding is further on. Once you have your permit, you can pick up the key at the police station in Gamboa at any time of day. They will make you leave something (your passport) as a deposit.

From the police station, which is on the left opposite a park, continue straight through Gamboa, and then when the paved road curves right, stay straight between the canal and a large pond that has a few Least Grebes and Common Moorhens. Capybara is also a possibility at night. Beyond the pond, turn right at a sign marking the beginning of Pipeline Road, and shortly you will come to the gate.

The road itself is mostly passible with 2WD as far as the Río Agua Salud, 17 km in, though in the rainy season there are mud holes that will be problematic. Trees across the road can also be a problem, hopefully on the way in, and not when you're trying to leave. The bridges (all in good condition now) over the various rivers are well-marked.

Much of the first 5 or 6 km is through broken second-growth, less interesting then what comes later, but still good for birds. Listen for mixed flocks. Shortly after the Río Limbo, a trail to the right leads to the Limbo Camp. This is a center for the Smithsonian Tropical Research Institute, and has some trails. Steer clear of active research projects.

More trails can be found at The Alamo (a concrete building with a framework of pipes), on the left shortly beyond Limbo. Look for the entrance to a network of trails on the left just beyond two old yellow gate posts. While the trails are less convenient for birding due to snakes and the possibility of getting lost, some species are not going to be seen from the road. The trails are also shady in the middle of the day.

After the Alamo the road is partly paved, and goes into hilly country. The forest here is better and your chances for most of the interesting birds also improves. The road passes over a number of attractive forest rivers, creeks really; these can be waded as an alternative way to enter the forest.

The Agua Salud is considered one of the better choices for wading, though if it is not accessible, one of the streams earlier on the road will do. Shorts and tennis shoes are probably the best gear, as there are pools that are thigh-deep in many places. Be sure to get your shoes dry as soon as possible, as these streams have a rather active bacterial flora. River wading is your only real chance for the Agami Heron and the Sunbittern, while some additional species are more likely to be found along the streams, especially during the dry season. The creeks are pleasant even when you're seeing nothing, which is a distinct possibility.

If you can acquire a tape, learn the calls of Spotted and especially Bicolored Antbirds, to facilitate the location of army ant swarms. The spectacle of a large antswarm is one of the most impressive sights of the tropics, and the accompanying birds here can include such specialties as Ocellated Antbird and Black-crowned Antpitta.

Pipeline is also good for mammals, with *Tamandua* anteater, Coati, Sloths, and Agouti among the possibilities. White-faced, Spider, and Howler Monkeys are also present, and the difference in behavior from areas with more human presence is readily evident.

## Birds of Pipeline Road

Great Tinamou
King Vulture
Double-toothed Kite
Plumbeous Hawk
White Hawk
Broad-winged Hawk
Black Hawk-Eagle
Slaty-backed Forest-Falcon
Crested Guan-r
Marbled Wood-Quail
Sunbittern
Short-billed Pigeon
Olive-backed Quail-Dove-r
Orange-chinned Parakeet
Red-lored Parrot
Rufous-vented Ground-Cuckoo-r
Crested Owl
Black-and-White Owl
Great Potoo
Band-rumped Swift
Rufous-breasted Hermit
Little Hermit
Violet-crowned Woodnymph
Purple-crowned Fairy
Violaceous Trogon
Slaty-tailed Trogon
Rufous Motmot
Green Kingfisher
Black-breasted Puffbird
Great Jacamar
Keel-billed Toucan

Agami Heron
Gray-headed Kite
Tiny Hawk
Semiplumbeous Hawk
Great Black-Hawk
Harpy Eagle-r
Barred Forest-Falcon
Collared Forest-Falcon
Great Curassow-r
Gray-necked Wood-Rail
Pale-vented Pigeon
Gray-chested Dove
Ruddy Quail-Dove
Blue-headed Parrot
Mealy Parrot
Vermiculated Screech-Owl
Mottled Owl
Short-tailed Nighthawk
Gray Potoo
Lesser Swallow-tailed Swift
Long-tailed Hermit
White-necked Jacobin
White-vented Plumeleteer
White-tailed Trogon
Black-throated Trogon
Blue-crowned Motmot
Broad-billed Motmot
White-necked Puffbird
White-whiskered Puffbird
Collared Aracari
Chestnut-mandibled Toucan

| | |
|---|---|
| Black-cheeked Woodpecker | Cinnamon Woodpecker |
| Lineated Woodpecker | Crimson-bellied Woodpecker-r |
| Crimson-crested Woodpecker | Slaty-winged Foliage-gleaner-r |
| Buff-throated Foliage-gleaner | Plain Xenops |
| Tawny-throated Leaftosser | Scaly-throated Leaftosser |
| Plain-brown Woodcreeper | Long-tailed Woodcreeper |
| Barred Woodcreeper | Buff-throated Woodcreeper |
| Black-striped Woodcreeper | Fasciated Antshrike |
| Great Antshrike | Slaty Antshrike |
| Spot-crowned Antvireo | Pygmy Antwren |
| Streaked Antwren | Checker-throated Antwren |
| White-flanked Antwren | Dot-winged Antwren |
| Dusky Antbird | Chestnut-backed Antbird |
| Dull-mantled Antbird-r | Spotted Antbird |
| Wing-banded Antbird | Bicolored Antbird |
| Ocellated Antbird | Black-faced Antthrush |
| Black-crowned Antpitta-r | Spectacled Antpitta |
| Forest Elaenia | Gray Elaenia-r |
| Greenish Elaenia | Ochre-bellied Flycatcher |
| Black-capped Pygmy-Tyrant | Southern Bentbill |
| Common Tody-Flycatcher | Brownish Flycatcher |
| Olivaceous Flatbill | Yellow-margined Flycatcher |
| Golden-crowned Spadebill | Royal Flycatcher |
| Ruddy-tailed Flycatcher | Sulphur-rumped Flycatcher |
| Black-tailed Flycatcher | migrant Empidonax |
| Long-tailed Tyrant | Bright-rumped Attila |
| Speckled Mourner-r | Rufous Mourner |
| Sirystes | Dusky-capped Flycatcher |
| White-ringed Flycatcher | Streaked Flycatcher |
| Masked Tityra | Rufous Piha |
| Blue Cotinga | Purple-throated Fruitcrow |
| Thrushlike Manakin | Broad-billed Sapoya |

Blue-crowned Manakin
Gray-breasted Martin
Black-bellied Wren
White-breasted Wood-Wren
Song Wren
Long-billed Gnatwren
Swainson's Thrush
Lesser Greenlet
Tennessee Warbler
Chestnut-sided Warbler
Northern Waterthrush
Mourning Warbler
Plain-colored Tanager
Blue Dacnis
Shining Honeycreeper
Fulvous-vented Euphonia
Gray-headed Tanager
Red-throated Ant-Tanager
Crimson-backed Tanager
Buff-throated Saltator
Slate-colored Grosbeak
Orange-billed Sparrow
Yellow-backed Oriole
Yellow-billed Cacique
Chestnut-headed Oropendola

Red-capped Manakin
White-thighed Swallow-r
Bay Wren
Southern Nightingale Wren
Tawny-faced Gnatwren
Tropical Gnatcatcher
Tawny-crowned Greenlet-r
Green Shrike-Vireo
Yellow Warbler
Bay-breasted Warbler
Kentucky Warbler
Bananaquit
Golden-masked Tanager
Green Honeycreeper
Red-legged Honeycreeper
White-vented Euphonia
White-shouldered Tanager
Summer Tanager
Dusky-faced Tanager
Black-headed Saltator
Blue-black Grosbeak
Black-striped Sparrow
Scarlet-rumped Cacique
Crested Oropendola

## The Atlantic Side of the Canal area

The Caribbean coast of central Panama has some excellent birding, with the disadvantage of visiting an area of such low socioeconomic development. The city of Colon is truly a sorry place, and has only two hotels that could be considered acceptable by most foreigners. DO NOT walk around Colon, even in broad daylight.

It is possible to stay in Panama City and drive over each day, but this obviously requires an early start. It is about an hour and forty minutes from the outskirts of Panama City to the start of Achiote Road; if you get caught at the locks, that could add considerably to the time needed to reach birding habitat.

Access to most of the birding areas requires that you cross the canal at the Gatun locks, since they are the only bridge on the Atlantic side. Arriving on the outskirts of Colon, you will come to a four-way stop, with the right turn going to France Field and the Free Zone, straight towards Colon itself, and left towards Gatun. Turn left, and shortly thereafter bear left on a road that will pass a small Chinese market before merging with the Bolivar Highway. After a couple of kilometers, follow signs to the right for Fort San Lorenzo.

The one-lane bridge across the bottom of the locks opens for the ships, so you may have to wait if one is passing. Eventually the gates will open, and you will get a green light.

Across the locks, the road goes right to Fort Sherman and Fort San Lorenzo, and more or less straight to Escobal Road. Escobal Road (signs for the Tarpon Club, overpriced food but convenient) has several side roads that are the main birding areas. The road crosses Gatun Dam, which has Red-breasted Blackbirds on its grassy slopes, before heading into the relatively-intact forest.

**Achiote Road**

Achiote Road leaves Escobal road at an unmarked intersection about 11 km after the Gatun Dam. Continuing habitat loss along the road means that the area is in decline, and directions based on the bridges might be invalidated by continuing road improvements. Loss of forest along the north side of the road has made the area rather open, so it gets hot early. Nonetheless, there are still several specialties to be looked for, and birding here is excellent early in the morning. As the day heats up, you can move on to S-9 and/or Tiger Trail.

The road is good for Hook-billed Kite and other raptors, Black-breasted and Pied Puffbirds, Spot-crowned Barbet, and tanager flocks. A trail to the right just before the third bridge leads to a small plantation of lowland coffee, with the overstory intact. This is currently the best area for White-headed Wren. There will be plenty of other action up and down the road.

**S-9 Road**
With continuing habitat loss on the Achiote Road, the S-9 Road has taken on more importance. It still passes through good forest and has a mostly closed canopy, so it is not as hot as the sun gets higher. The turn is marked with an Army sign where it leaves Escobal Road, about 4 km after the dam. The birds here are not much different from Achiote, with perhaps less variety.

**Tiger Trail**
Just past the Gatun dam, a gravel road goes right to the Tiger Trail. The early parts of the trail are hard to follow, but shortly it settles onto an old railway embankment and continues under a low canopy. The open understory allows for pursuit of forest-floor species such as leaf-tossers and quail-doves.

**Fort San Lorenzo**
The area around the ruins of the old Fort San Lorenzo is one of the most attractive in Central Panama and has good birding too. Leaving the Gatun locks, bear generally right and go towards Fort Sherman on the good paved road.

At the entrance to the facility, a guard will ask for i.d. and possibly proof of car insurance. (Note: access may change when the fort is handed over to the Panamanians.) Continue through the base following signs for Fort San Lorenzo. Soon the road turns to gravel and enters good habitat; the best is towards the end of the road.

There are quite a few side roads on the way out to the old fort, and any not posted by the military can be good birding. Just before the coast, a road to the left goes to the Chagres River mouth, and continues inland through an interesting swamp forest. There are trails off the end of this road as well.

**Mangroves**
There are some interesting mangroves on the way to Fort Sherman, but the better habitat is on the other side of the canal. To reach the mangrove along Galeta Road, bear right at the intersection as you enter Colon. After passing the airport, the free zone, and a large lot of parked cars, the road will bear to the right towards the Galeta Naval Station. Don't approach the gates of the station; turn around once they come into view. The mangroves here have a number of associated bird species, including Common Black-Hawk, Greater Ani, Black-tailed Trogon, six kingfishers, Straight-billed and Streak-headed Woodcreepers, Lesser Kiskadee, Rusty-margined Flycatcher, and "Mangrove" Warbler.

## Birds of the Atlantic Side

Great Tinamou
Rufescent Tiger-Heron
Gray-headed Kite
Double-toothed Kite
Semiplumbeous Hawk
White Hawk
Broad-winged Hawk
Black Hawk-Eagle
Slaty-backed Forest-Falcon
Peregrine Falcon
Gray-necked Wood-Rail
Pale-vented Pigeon
White-tipped Dove
Ruddy Quail-Dove
Blue-headed Parrot
Mealy Amazon
Crested Owl
Black-and-White Owl
Gray Potoo
Lesser Swallow-tailed Swift
Band-tailed Barbthroat
Little Hermit
White-necked Jacobin
Rufous-crested Coquette-r
Violet-bellied Hummingbird
Blue-chested Hummingbird
White-tailed Trogon
Black-throated Trogon
Slaty-tailed Trogon
Rufous Motmot
Ringed Kingfisher

Little Tinamou
King Vulture
Hook-billed Kite
Plumbeous Hawk
Common Black-Hawk
Great Black-Hawk
Short-tailed Hawk
Barred Forest-Falcon
Collared Forest-Falcon
Marbled Wood-Quail
Scaled Pigeon
Short-billed Pigeon
Gray-chested Dove
Orange-chinned Parakeet
Red-lored Amazon
Vermiculated Screech-Owl
Mottled Owl
Great Potoo
Band-rumped Swift
Rufous-breasted Hermit
Long-tailed Hermit
Scaly-breasted Hummingbird
Black-throated Mango
Violet-crowned Woodnymph
Sapphire-throated Hummingbird
Purple-crowned Fairy
Violaceous Trogon
Black-tailed Trogon
Blue-crowned Motmot
Broad-billed Motmot
Amazon Kingfisher

| | |
|---|---|
| Green Kingfisher | White-necked Puffbird |
| Black-breasted Puffbird | Pied Puffbird |
| White-whiskered Puffbird | Collared Aracari |
| Keel-billed Toucan | Chestnut-mandibled Toucan |
| Black-cheeked Woodpecker | Cinnamon Woodpecker |
| Lineated Woodpecker | Crimson-crested Woodpecker |
| Buff-throated Foliage-gleaner | Plain Xenops |
| Tawny-throated Leaftosser | Scaly-throated Leaftosser |
| Plain-brown Woodcreeper | Long-tailed Woodcreeper |
| Barred Woodcreeper | Buff-throated Woodcreeper |
| Black-striped Woodcreeper | Fasciated Antshrike |
| Great Antshrike | Slaty Antshrike |
| Spot-crowned Antvireo | Pygmy Antwren |
| Streaked Antwren | Checker-throated Antwren |
| White-flanked Antwren | Dot-winged Antwren |
| Dusky Antbird | Jet Antbird |
| Bare-crowned Antbird | Chestnut-backed Antbird |
| Spotted Antbird | Bicolored Antbird |
| Ocellated Antbird | Black-faced Antthrush |
| Black-crowned Antpitta-r | Spectacled Antpitta |
| Brown-capped Tyrannulet | Greenish Elaenia |
| Ochre-bellied Flycatcher | Southern Bentbill |
| Common Tody-Flycatcher | Black-headed Tody-Flycatcher |
| Olivaceous Flatbill | Yellow-margined Flycatcher |
| Golden-crowned Spadebill | Royal Flycatcher |
| Ruddy-tailed Flycatcher | Black-tailed Flycatcher |
| migrant Empidonax | Long-tailed Tyrant |
| Bright-rumped Attila | Rufous Mourner |
| Dusky-capped Flycatcher | Panama Flycatcher |
| White-ringed Flycatcher | Streaked Flycatcher |
| White-winged Becard | Masked Tityra |
| Purple-throated Fruitcrow | Thrushlike Manakin |

| | |
|---|---|
| Golden-collared Manakin | Blue-crowned Manakin |
| Red-capped Manakin | Gray-breasted Martin |
| Mangrove Swallow | White-thighed Swallow-r |
| Southern Rough-winged Swallow | migrant swallows |
| White-headed Wren | Black-bellied Wren |
| Bay Wren | Buff-breasted Wren |
| White-breasted Wood-Wren | Southern Nightingale Wren |
| Song Wren | Long-billed Gnatwren |
| Tropical Gnatcatcher | Swainson's Thrush |
| Lesser Greenlet | Green Shrike-Vireo |
| Tennessee Warbler | Yellow Warbler |
| Chestnut-sided Warbler | Bay-breasted Warbler |
| Northern Waterthrush | Kentucky Warbler |
| Mourning Warbler | Bananaquit |
| Plain-colored Tanager | Bay-headed Tanager |
| Golden-masked Tanager | Scarlet-thighed Dacnis |
| Blue Dacnis | Green Honeycreeper |
| Shining Honeycreeper | Red-legged Honeycreeper |
| Thick-billed Euphonia | Fulvous-vented Euphonia |
| White-vented Euphonia | Gray-headed Tanager |
| Sulphur-rumped Tanager | White-shouldered Tanager |
| Red-throated Ant-Tanager | Summer Tanager |
| Crimson-backed Tanager | Flame-rumped Tanager |
| Dusky-faced Tanager | Buff-throated Saltator |
| Black-headed Saltator | Slate-colored Grosbeak |
| Blue-black Grosbeak | Orange-billed Sparrow |
| Black-striped Sparrow | Yellow-backed Oriole |
| Scarlet-rumped Cacique | Yellow-billed Cacique |
| Crested Oropendola | Chestnut-headed Oropendola |

## Tocumen Marsh

Though most of the Tocumen marsh has been converted to rice fields, there still is good birding in the area. A number of South American species reach what is essentially the northern extreme of their ranges here.

To reach the marsh, turn left at the Riande Airport Hotel. After about 7 km there is a junction with a police post. Turn right here, and then left when the road forks about a kilometer later. Just past a gas station, about 2.5 km from the fork, a gravel road turn left towards the farm. At the maintenance buildings, ask permission to enter and try to ascertain that the gate will not be closed with you inside.

Mostly the roads in the rice plantation are rough but passible. After a rain they can be slippery, and shouldn't be attempted without 4WD.

At the shop, the road forks; to the right leads to most of the better birding. Explore the dikes, looking for herons and other water birds in the flooded areas and Southern Lapwing in newly-turned fields. Continuing straight at the fork by the shop takes you to more rice fields; the next road to the right also leads to good habitat. If possible, follow both these roads to the end in order to access the remaining natural marsh areas, where there might be night-herons or Bare-throated Tiger-Heron.

## Birds of Tocumen

| | |
|---|---|
| Anhinga | Bare-throated Tiger-Heron |
| Great Blue Heron | Cocoi Heron |
| Great Egret | Snowy Egret |
| Little Blue Heron | Green Heron |
| Striated Heron | Capped Heron |
| Black-crowned Night-Heron | Boat-billed Heron |
| Least Bittern | Wood Stork |
| White Ibis | Blue-winged Teal |
| Lesser Yellow-headed Vulture | White-tailed Kite |
| Pearl Kite | Crested Caracara |
| Purple Gallinule | Southern Lapwing |
| Wattled Jacana | migrant shorebirds |

| | |
|---|---|
| Ruddy Ground-Dove | Blue Ground-Dove |
| White-tipped Dove | Yellow-crowned Parrot |
| Little Cuckoo | Greater Ani |
| Smooth-billed Ani | Barn Owl |
| Black-throated Mango | Ringed Kingfisher |
| Amazon Kingfisher | Green Kingfisher |
| Green-and-Rufous Kingfisher | American Pygmy Kingfisher |
| Red-crowned Woodpecker | Pale-breasted Spinetail |
| Barred Antshrike | Southern Beardless-Tyrannulet |
| Mouse-colored Tyrannulet | Common Tody-Flycatcher |
| Pied Water-Tyrant | Panama Flycatcher |
| Lesser Kiskadee | Rusty-margined Flycatcher |
| Fork-tailed Flycatcher | Gray-breasted Martin |
| Barn Swallow | House Wren |
| Streaked Saltator | Black-striped Sparrow |
| Blue-black Grassquit | Variable Seedeater |
| Yellow-bellied Seedeater | Red-breasted Blackbird |
| Eastern Meadowlark | Lesser Goldfinch |

## Cerro Azul/Jefe

Some specialties of the eastern Panama foothills can be found close to the city at Cerro Azul and Cerro Jefe. Here you can escape from the hot lowlands for a bit of mountain birding. The weather can be difficult, so come prepared for mist and cold fog.

From the Riande Airport Hotel, go towards Chepo for 7 km to a transit police station. Here, turn left, and after a couple of kilometers, left again just after a Chinese pavilion. Since this may not be obvious in the dark, look also for signs for the various housing developments on Cerro Azul. The road climbs through mixed woodland that has some interesting birding. In the residential neighborhoods there are still roads with enough habitat to support mixed flocks of tanagers, etc.

Eventually the road turns to dirt but continues to the towers on

Cerro Jefe. The last 2 or 3 km is too rough for a passenger car. In the elfin forest around the summit (check behind the towers as well), look for Violet-capped Hummingbird, Tacarcuna Bush-Tanager, and Black-headed Brush-Finch.

## Birds of Cerro Azul/Jefe

Scaled Pigeon
Blue-fronted Parrotlet-r
Rufous Nightjar
Violet-capped Hummingbird
Keel-billed Toucan
Olivaceous Woodcreeper
Ochre-bellied Flycatcher
Olivaceous Flatbill
Dusky-capped Flycatcher
Gray-breasted Martin
S. Nightingale Wren
Yellow-throated Vireo
Black-and-White Warbler
Bay-breasted Warbler
Emerald Tanager
Bay-headed Tanager
Blue Dacnis
White-vented Euphonia
Olive Tanager
Tacarcuna Bush-Tanager
Black-headed Brush-Finch
Yellow-faced Grassquit

Orange-chinned Parakeet
Pauraque
Green Hermit
Violet-headed Hummingbird
Black-cheeked Woodpecker
Black-headed Antthrush-r
Scale-crested Pygmy-Tyrant
Ruddy-tailed Flycatcher
Streaked Flycatcher
Gray-breasted Wood-Wren
Tropical Mockingbird
Lesser Greenlet
Chestnut-sided Warbler
Plain-colored Tanager
Speckled Tanager
Golden-hooded Tanager
Green Honeycreeper
Fulvous-vented Euphonia
Black-and-Yellow Tanager
Blue-black Grosbeak
Variable Seedeater

# Eastern Panama

The spectacular birding of the Darién is mostly out-of-reach for those unable to mount an expedition, except for the areas below. Most of the numerous specialties can be found in either Cana or El Real.

## Cana

The Darién is most easily sampled by chartering a plane to visit the old mining station at Cana. It goes without saying that this is expensive; a group of five is a practical minimum. This is the one spot in Central America where you might be better off on a commercial tour.

The best time to visit is during the dry season from late December to April. At other times the grass airstrip may be too wet to use. Keep in mind that tour companies target this time of year too, and the facility may be booked well in advance.

The station is run by ANCON, a Panamanian non-governmental conservation organization. Facilities are primitive, but there are flush toilets and a generator that runs a few hours each night. ANCON will only deal with a Panamanian travel agency, so you will need to set up a package with someone in Panama. Allow lots of time for arrangements.

A guide is another requirement. Wilberto Martinez is one who has a good reputation. He works with Pezantes Tours (Apt. 55-0716, Paitilla, Panama, Republic of Panama; (507) 263-7577, fax 263-7860).

More economical possibilities can be arranged with Yenia Mendoza of Agencia de Viajes Jazmine, (Calle 46, Edificio Miramar, Apartado 4620, Panama 5, Republic of Panama, tel. (507) 223-0179, fax 264-0309. Expect to pay at least $1,000 per person for 5 days all inclusive; prices will vary depending on group size, etc.

The birding at Cana is excellent, with a variety of difficult-to-find species. For a complete list, see the checklist in this book.

## El Real

Some of the lowland Darién specialties not found at Cana can be seen at the town of El Real. There are commercial flights three times a week for about $65 round trip, and the hotel in town should be adequate. Birding for Red-billed Scythebill, Black-capped Donocobius, Black Oropendola, etc. is around the town. Take precautions against malaria here.

# Western Panama

# Western Foothills

Several locations in the mountains between the canal and Chiriquí offer interesting birding, if only to break up a trip to the west. While lacking the specialties of the Costa Rica-Chiriquí mountains, these areas have a good variety of species not found in the Central Panama lowlands. The Aguadulce mud flats are included here due to their geographic position.

The foothill locations detailed below, particularly El Copé and Santa Fé, suffer from consistently bad weather. Your best chance of sun is in March, but mud is likely year-round. The birds take the rain, wind, and mist in stride, so you should too.

## Cerro Campana

Cerro Campana is the nearest of the western foothill birding areas, about 70 minutes from the bridge over the canal. The turn to the national park is about 6 km west of Capira, and is well-marked from the Panama City direction. The road up the mountain is in poor condition, but should be passible with a 2WD vehicle.

At the park station a few kilometers along the road, stop and check in, then look for Wedge-tailed Grass-Finch on the barren slopes nearby. The main birding area is near the top of the mountain, though there is some broken habitat along the road. Where the road forks, bear right past the microwave towers and a couple of houses into the montane forest. The park has a nature trail set up here; inquire at the station for details. Here a network of roads allows good coverage of the area, which has Black-crowned Antpitta and flocks of tanagers.

Economical accommodation is available at Richard's Place (no sign), at the very top of the mountain. Where the road forks just before the main microwave facility, bear left to the end of the pavement. Here you'll see a complex of buildings with a couple of radio towers; the building nearest the road has a facade of concrete steps. No food is available anywhere on Cerro Campana.

## Birds of Cerro Campana

| | |
|---|---|
| Swallow-tailed Kite | White-tailed Hawk-r |
| Mourning Dove | Purplish-backed Quail-Dove-r |
| Pauraque | Little Hermit |
| Snowy-bellied Hummingbird | Collared Aracari |
| Keel-billed Toucan | Orange-bellied Trogon |
| Plain Xenops | Slaty Antshrike |
| Plain Antvireo | Black-faced Antthrush |
| Black-headed Antthrush-r | Black-crowned Antpitta-r |
| Ochre-bellied Flycatcher | Scale-crested Pygmy-Tyrant |
| Dusky-capped Flycatcher | White-ruffed Manakin |
| Blue-and-White Swallow | Gray-breasted Wood-Wren |
| White-breasted Wood-Wren | Song Wren |
| White-throated Thrush | Lesser Greenlet |
| Silver-throated Tanager | Bay-headed Tanager |
| Green Honeycreeper | Shining Honeycreeper |
| Thick-billed Euphonia | Tawny-capped Euphonia |
| Olive Tanager | Tawny-crested Tanager |
| Red-crowned Ant-Tanager | Summer Tanager |
| Hepatic Tanager | Black-and-Yellow Tanager |
| Blue-black Grosbeak | Chestnut-capped Brush-Finch |
| Black-striped Sparrow | Yellow-faced Grassquit |
| Wedge-tailed Grass-Finch | Rufous-collared Sparrow |

## El Copé

The forests above the town of El Copé are very attractive, if the clouds allow you to see them. The habitat on the Caribbean slope is extensive.

There are no accommodations closer then Penonomé, a little more than an hour's drive from the birding area. There are a couple of 24-hour restaurants along the highway in and near Penonomé.

The road to the town of El Copé leaves the Pan-American highway about 20 km west of Penonomé. It is well-marked with a sign for Parque

Nacional El Copé. The road winds 27 km up to the town. As you reach the houses, bear right on a street marked with a one-way sign. Don't be distracted by a sign to the INRENARE park headquarters. Stay to the right, back down the hill, without going into the main section of the village. If in doubt ask for the "Escuela de Barrigón." The last 4 km or so of the road requires high clearance, though you might make it in a car with a great deal of care.

The road essentially becomes undrivable at a large clearing with some decomposing logging equipment, with a fork shortly after. The right fork has better birding, continuing along the Caribbean slope for several kilometers. The left fork climbs and generally stays on the Pacific side.

## Birds of El Copé

| | |
|---|---|
| Little Tinamou | Swallow-tailed Kite |
| Double-toothed Kite | Barred Hawk |
| Short-tailed Hawk | Barred Forest-Falcon |
| Gray-headed Chachalaca | Black Guan-r |
| Scaled Pigeon | Short-billed Pigeon |
| Purplish-backed Quail-Dove-r | Red-fronted Parrotlet-r |
| Blue-headed Parrot | Ferruginous Pygmy-Owl |
| Band-tailed Barbthroat | Little Hermit |
| Long-tailed Hermit | Green Hermit |
| White-tipped Sicklebill | Brown Violetear-r |
| Violet-headed Hummingbird | Green Thorntail |
| Snowcap | Variable Mountain-gem |
| Orange-bellied Trogon | Slaty-tailed Trogon |
| Lattice-tailed Trogon | Rufous Motmot |
| Red-headed Barbet | Emerald Toucanet |
| Yellow-eared Toucanet | Golden-olive Woodpecker |
| Cinnamon Woodpecker | Spotted Barbtail |
| Slaty-winged Foliage-gleaner | Plain Xenops |
| Plain-brown Woodcreeper | Spotted Woodcreeper |

Brown-billed Scythebill
Plain Antvireo
Slaty Antwren
Immaculate Antbird
Black-crowned Antpitta-r
Paltry Tyrannulet
Ochre-bellied Flycatcher
Yellow-margined Flycatcher
Sulphur-rumped Flycatcher
Yellow-bellied Flycatcher
Speckled Mourner-r
Dusky-capped Flycatcher
Three-wattled Bellbird
White-ruffed Manakin
Black-chested Jay
Stripe-breasted Wren
White-breasted Wood-Wren
S. Nightingale Wren
Long-billed Gnatwren
Slaty-backed Nightingale-Thrush
Pale-vented Thrush
White-throated Thrush
Green Shrike-Vireo
Tennessee Warbler
Blackburnian Warbler
Black-and-White Warbler
Louisiana Waterthrush
Slate-throated Whitestart
Bananaquit
Emerald Tanager
Speckled Tanager
Scarlet-thighed Dacnis
Shining Honeycreeper

Russet Antshrike
Spot-crowned Antvireo
Chestnut-backed Antbird
Black-faced Antthrush
Spectacled Antpitta
Olive-striped Flycatcher
Slaty-capped Flycatcher
White-throated Spadebill
Tufted Flycatcher
Bright-rumped Attila
Rufous Mourner
Cinnamon Becard
Golden-collared Manakin
Blue-and-White Swallow
Bay Wren
Rufous-breasted Wren
Gray-breasted Wood-Wren
Tawny-faced Gnatwren
Black-faced Solitaire
Swainson's Thrush
Clay-colored Thrush
Lesser Greenlet
Golden-winged Warbler
Chestnut-sided Warbler
Bay-breasted Warbler
American Redstart
Mourning Warbler
Buff-rumped Warbler
Plain-colored Tanager
Silver-throated Tanager
Bay-headed Tanager
Green Honeycreeper
Red-legged Honeycreeper

| | |
|---|---|
| White-vented Euphonia | Tawny-capped Euphonia |
| Hepatic Tanager | Summer Tanager |
| Blue-and-Gold Tanager | Olive Tanager |
| White-throated Shrike-Tanager | White-lined Tanager |
| Red-crowned Ant-Tanager | Red-throated Ant-Tanager |
| Tawny-crested Tanager | Dusky-faced Tanager |
| Common Bush-Tanager | Yellow-throated Bush-Tanager |
| Black-and-Yellow Tanager | Streaked Saltator |
| Buff-throated Saltator | Slate-colored Grosbeak |
| Black-faced Grosbeak | Chestnut-capped Brush-Finch |
| Orange-billed Sparrow | Black-striped Sparrow |
| Variable Seedeater | Blue Seedeater-r |
| Scarlet-rumped Cacique | |

## Aguadulce

The tidal flats near the town of Aguadulce have a few shorebirds as well as some mangrove specialties. The ponds are generally better at high tide, as there are extensive mudflats nearby exposed at low tide.

The birding area is only a few kilometers off the main highway. There are hotels and restaurants in Aguadulce.

Coming from Panamá, turn left at the blinking light at the main intersection in Aguadulce, just past the Hotel Intercontinental. After a few blocks turn left at a T intersection, then right just before a small plaza with a statue. Upon reaching the main plaza, go around to the far corner and exit the square between the Agrocomercial del Pacífico and Materiales La Economía (the other streets here are one-way against you). Immediately bear right twice around the block and then left. When this road forks, bear left and cross a bridge. At the next intersection bear left past the peach-colored park benches. A sports complex will appear shortly on the right. After 3 km on this road, there are some ponds on both sides of the road. To reach the ocean continue another 4 km across the salt works.

The ponds often have shorebirds, and the mangroves have Mangrove Black-Hawk, Northern Scrub-Flycatcher, and "Mangrove" Yellow and Prothonotary Warblers.

## Birds of the Aguadulce tidal flats

| | |
|---|---|
| herons & egrets | Mangrove Black-Hawk |
| Yellow-headed Caracara | migrant shorebirds |
| White-winged Dove | Plain-breasted Ground-Dove |
| Orange-chinned Parakeet | Yellow-crowned Parrot |
| Striped Cuckoo | Green Kingfisher |
| N. Scrub-Flycatcher | Fork-tailed Flycatcher |
| Mangrove Warbler | Prothonotary Warbler |
| Northern Waterthrush | |

## Santa Fé

The town of Santa Fé is 57 km north of Santiago, near the continental divide, with access to good habitat. Much of the birdlife here has a distinctly Caribbean flavor, despite the fact that the site is on the Pacific slope. This is the best place in Panama to see the Lattice-tailed Trogon, among other species.

There is one small hotel on the road into town, and a few restaurants in Santa Fé. There are numerous facilities in Santiago, slightly more than an hour away.

Entering town, bear left at each of two forks. After the pavement ends, take the first left towards the Alto de Piedra forestry school. After about 4 km you will see the school complex; shortly after that there is a fork in the road. The left fork goes 1 km or so into habitat that is mostly second-growth in various stages of regeneration. The better birding is along the right fork. Here the road deteriorates rapidly, and is no longer passable even with 4WD after it starts down the first ridge.

Nonetheless, it is only about 500 meters from that point to where the good habitat begins, and road continues for miles into an Indian reservation.

## Birds of Santa Fé

| | |
|---|---|
| Great Tinamou | Little Tinamou |
| Swallow-tailed Kite | Barred Hawk |
| White Hawk | Solitary Eagle-r |
| Short-tailed Hawk | Black Hawk-Eagle |
| Gray-headed Chachalaca | Sunbittern-r |
| Short-billed Pigeon | Brown-hooded Parrot |
| Squirrel Cuckoo | Pauraque |
| White-collared Swift | Vaux's Swift |
| Band-rumped Swift | Band-tailed Barbthroat |
| Green Hermit | Long-tailed Hermit |
| Little Hermit | White-necked Jacobin |
| Brown Violetear-r | Violet-headed Hummingbird |
| Snowy-bellied Hummingbird | White-tailed Emerald |
| Snowcap | Purple-crowned Fairy |
| Rufous-tailed Hummingbird | Bronze-tailed Plumeleteer |
| White-bellied Mountain-gem | Variable Mountain-gem |
| Orange-bellied Trogon | Lattice-tailed Trogon |
| Rufous Motmot | Emerald Toucanet |
| Yellow-eared Toucanet | Keel-billed Toucan |
| Chestnut-mandibled Toucan | Smoky-brown Woodpecker |
| Rufous-winged Woodpecker-r | Golden-olive Woodpecker |
| Cinnamon Woodpecker | Spotted Barbtail |
| Striped Woodhaunter | Lineated Foliage-gleaner |
| Plain Xenops | Plain-brown Woodcreeper |
| Spotted Woodcreeper | Brown-billed Scythebill |
| Fasciated Antshrike | Russet Antshrike |
| Plain Antvireo | White-flanked Antwren |
| Slaty Antwren | Dot-winged Antwren |
| Chestnut-backed Antbird | Dull-mantled Antbird-r |
| Immaculate Antbird | Bicolored Antbird |
| Black-crowned Antpitta-r | Spectacled Antpitta |

| | |
|---|---|
| Paltry Tyrannulet | S. Beardless Tyrannulet |
| Olive-striped Flycatcher | Slaty-capped Flycatcher |
| Rufous-browed Tyrannulet-r | Scale-crested Pygmy-Tyrant |
| Brownish Flycatcher-r | Eye-ringed Flatbill |
| Olivaceous Flatbill | White-throated Spadebill |
| Ruddy-tailed Flycatcher | Black-tailed Flycatcher |
| Tufted Flycatcher | Acadian Flycatcher |
| Long-tailed Tyrant | Bright-rumped Attila |
| Rufous Mourner | Dusky-capped Flycatcher |
| Cinnamon Becard | Purple-throated Fruitcrow |
| Thrushlike Manakin | Golden-collared Manakin |
| White-ruffed Manakin | Lance-tailed Manakin |
| Red-capped Manakin | Black-chested Jay |
| Band-backed Wren | Bay Wren |
| Stripe-breasted Wren | Rufous-breasted Wren |
| Rufous-and-White Wren | White-breasted Wood-Wren |
| Gray-breasted Wood-Wren | S. Nightingale Wren |
| Tawny-faced Gnatwren | Long-billed Gnatwren |
| Tropical Gnatcatcher | Black-faced Solitaire |
| Orange-billed Nightingale-Thrush | Slaty-backed Nightingale-Thrush |
| Swainson's Thrush | Pale-vented Thrush |
| Clay-colored Thrush | White-throated Thrush |
| Tawny-crowned Greenlet | Lesser Greenlet |
| Golden-winged Warbler | Tropical Parula |
| Blackburnian Warbler | Mourning Warbler |
| Slate-throated Whitestart | Golden-crowned Warbler |
| Three-striped Warbler | Buff-rumped Warbler |
| Bananaquit | Emerald Tanager |
| Silver-throated Tanager | Speckled Tanager |
| Bay-headed Tanager | Green Honeycreeper |
| Shining Honeycreeper | Thick-billed Euphonia |
| Blue-hooded Euphonia | White-vented Euphonia |
| Tawny-capped Euphonia | Blue-and-Gold Tanager |

| | |
|---|---|
| Olive Tanager | White-throated Shrike-Tanager |
| Tawny-crested Tanager | Red-crowned Ant-Tanager |
| Red-throated Ant-Tanager | Hepatic Tanager |
| Flame-rumped Tanager | Dusky-faced Tanager |
| Common Bush-Tanager | Yellow-throated Bush-Tanager |
| Black-and-Yellow Tanager | Buff-throated Saltator |
| Slate-colored Grosbeak | Blue-black Grosbeak |
| Sooty-faced Finch-r | Orange-billed Sparrow |
| Black-striped Sparrow | Variable Seedeater |
| Lesser Seed-Finch | Yellow-faced Grassquit |
| Yellow-billed Cacique | Scarlet-rumped Cacique |
| Crested Oropendola | Chestnut-headed Oropendola |

# Chiriquí

The western Chiriquí highlands around the extinct Volcán Barú have many bird species otherwise restricted to Costa Rica. The pleasant weather makes the area a popular vacation spot for Panamanians, so there are good accommodations and other facilities.

The area around the Fortuna reservoir also provides good birding, and has the advantage of easy access to the Caribbean slope foothills of Bocas del Toro province.

Farther east but still in Chiriquí, the Cerro Colorado area has two very localized species. Access here is more difficult than in the western part of the province.

David, Panama's third largest city, is about 6 hours non-stop from Panama City. The highway west from the capital is mostly in good condition. The lowlands of Chiriquí and Veraguas have been almost entirely deforested, severely reducing the Panama distribution of a number of species, but offering the chance of Savannah Hawk and Yellow-crowned Parrot. A small area of remnant woodland near Chorcha Abajo can provide a few species of the lowland Chiriquí avifauna.

It is also possible to fly to David and rent a car to avoid the long drive, though by the time you go to that much trouble you might as well fly to Costa Rica. Plane fares to David are about $100 round trip. Keep in mind that domestic flights leave from a different airport in Panama

City, so connections to your international flights are difficult.

The towns of Boquete and Volcán are the obvious bases for birding the Volcán Barú area, both at a refreshing 1,000 meters or so. There are plenty of hotels and restaurants in both towns as well as Cerro Punta.

## Boquete

This pleasant town is about 35 km from David on the lower edge of the highlands. Birding here is not generally as good as on the west side of the volcano, but the road to the top of Volcán Barú starts near town, providing access to the highest habitats.

A well-known birding area near Boquete is the Finca Lérida, owned by the same family as the Hotel Panamonte in town. I was told that entrance costs $25 per day.

The road to the top of Volcán Barú starts as Calle 2a Norte (no signs) in Boquete, the first left past the church on the main drag in town. The road leaves town to the west past a sign for Volcán Barú National Park. Stay on the asphalt straight to a park entrance station, and then continue up the main road to the top. The road is very rough and loose, and some spots are difficult even with 4WD.

The habitat here is patchy until you are well up the mountain, when good montane forest with Quetzals is encountered. Farther up, the forest gets brushier with different birds, such as Timberline Wren, Sooty Thrush, and many Black-billed Nightingale-Thrushes. At the very top there is low scrub that has Volcano Junco.

## Volcán

This town is about the same elevation as Boquete on the other side of the volcano. There is no road connecting them; you must descend to David, go towards Costa Rica, and then turn at the town of Concepción. The highway up to Volcán is well-marked coming from David, less so if you are arriving from the Costa Rican border. Look for signs for the Hotel Bambito. Volcán is slightly less than an hour from David.

One birding area is the second-growth forest around the Volcán Lakes west of town. To reach the lakes, take the left fork at the Guardia post in Volcán and then a left on the fifth street from the fork, just after a small video rental place. Jog right at a T intersection and stay on the pavement across the airstrip. Straight across the runway, a gravel road continues for about a kilometer to the lakes. The best access to the lakes is through a black metal gate on the left fork. The road is rough in places, but should be possible with care in an ordinary car. The lakes themselves have a few herons, gallinules, etc. while the forest around them is good for mixed flocks. Keep an eye on your car.

The right fork in Volcán goes to Bambito and Cerro Punta. Check along the Río Chiriquí Viejo for North American Dipper, Black Phoebe, and Torrent Tyrannulet.

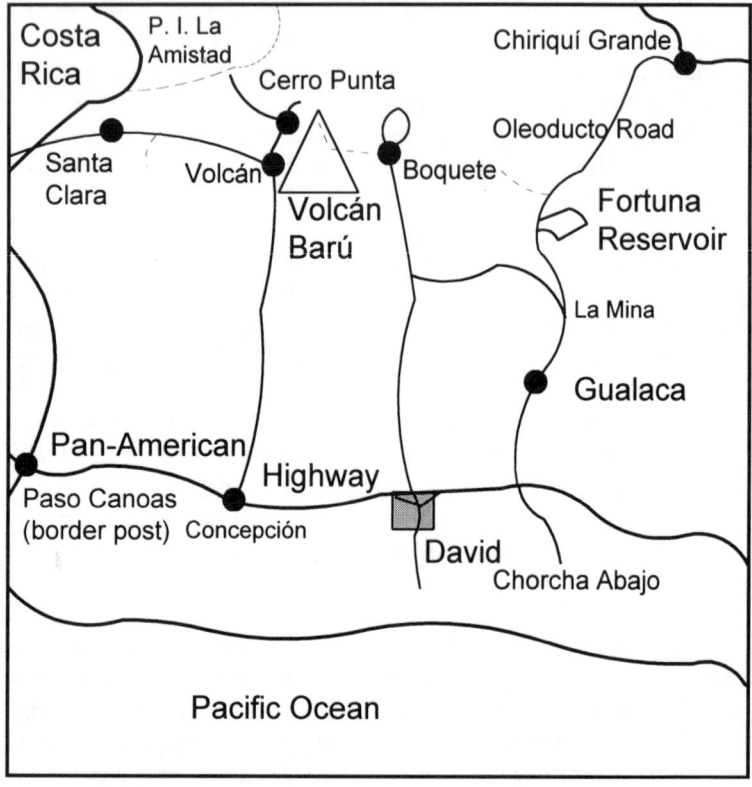

**Cerro Punta**

Somewhat higher-elevation forest can be reached from Cerro Punta. Stay on the main road, which about 1 km after Cerro Punta drops down to cross a bridge and then forks, with the right fork paved. Stay right until the road starts up a steep hill, and then walk the road up to the ridge. This road is good for higher-elevation forest species, while a trail continues on to Boquete.

The house at the end of the pavement belongs to the Fern family, who work as guides to the area. Their rate is $35 per day ₁ group of fewer than five people.

**Parque Internacional La Amistad**
From the village of Cerro Punta, a road to the left at the Agroblas fertilizer company goes to the Las Nubes park guard station. After about a kilometer, take a left at a signed crossroads. About 4 km from Cerro Punta the pavement ends, and the road gets rough in places for the last 3 km. If you make it to the third bridge after the pavement ends, it may be better to walk the last few hundred meters through interesting habitat to the park station. There are trails into the park from there.

**Santa Clara**
The left fork at the police station in Volcán (mentioned above) continues through broken habitat to the village of Santa Clara. About 21 kilometers from the fork in Volcán, a road to the left provides access to some interesting habitat. To facilitate recognition, the road takes off from the outside of a curve opposite another dirt road to the right, and follows a cutbank along the top of a small ridge for a hundred meters or so before starting down the hill. Any car without 4WD should not go onto the steep part. In this foothill forest, there is a chance for Fiery-billed Aracari, Spot-crowned Euphonia, and Scarlet-thighed Dacnis.

## Birds of the Western Chiriquí Highlands

| | |
|---|---|
| Black Guan | Spotted Wood-Quail |
| Scaled Pigeon | Band-tailed Pigeon |
| Chiriquí Quail-Dove | Buff-fronted Quail-Dove |
| Sulphur-winged Parakeet | Squirrel Cuckoo |
| Mottled Owl | Dusky Nightjar |
| White-collared Swift | Vaux's Swift |
| Green Hermit | Violet Sabrewing |
| Green Violetear | Fiery-throated Hummingbird |
| White-tailed Emerald | Variable Mountain-gem |
| Volcano Hummingbird | Scintillant Hummingbird |

| | |
|---|---|
| Collared Trogon | Resplendent Quetzal |
| Blue-crowned Motmot | Emerald Toucanet |
| Fiery-billed Aracari | Acorn Woodpecker |
| Hairy Woodpecker | Pale-breasted Spinetail |
| Slaty Spinetail | Red-faced Spinetail |
| Ruddy Treerunner | Buffy Tuftedcheek |
| Spectacled Foliage-gleaner | Ruddy Foliage-gleaner |
| Streak-breasted Treehunter | Streak-headed Woodcreeper |
| Spot-crowned Woodcreeper | Silvery-fronted Tapaculo |
| Paltry Tyrannulet | Mountain Elaenia |
| Torrent Tyrannulet | Tufted Flycatcher |
| Dark Pewee | Black-capped Flycatcher |
| Black Phoebe | Dusky-capped Flycatcher |
| Golden-bellied Flycatcher | Streaked Flycatcher |
| Piratic Flycatcher | Rose-throated Becard-r |
| Masked Tityra | Three-wattled Bellbird |
| Gray-breasted Martin | Blue-and-White Swallow |
| Rough-winged Swallows | Black-chested Jay |
| Rufous-breasted Wren | Plain Wren |
| Ochraceous Wren | Timberline Wren |
| Gray-breasted Wood-Wren | N. Am. Dipper |
| Black-faced Solitaire | Black-billed Nightingale-Thrush |
| Ruddy-capped Nightingale-Thrush | Sooty Thrush |
| Mountain Thrush | Clay-colored Thrush |
| Black-and-Yellow Silky-Flycatcher | Long-tailed Silky-Flycatcher |
| Yellow-winged Vireo | Brown-capped Vireo |
| Lesser Greenlet | Rufous-browed Peppershrike |
| Golden-winged Warbler | Tennessee Warbler |
| Tropical Parula | Flame-throated Warbler |
| Chestnut-sided Warbler | Black-throated Green Warbler |
| Blackburnian Warbler | Mourning Warbler |
| Masked Yellowthroat | Wilson's Warbler |

| | |
|---|---|
| Slate-throated Whitestart | Collared Whitestart |
| Golden-crowned Warbler | Rufous-capped Warbler |
| Black-cheeked Warbler | Silver-throated Tanager |
| Bay-headed Tanager | Golden-hooded Tanager |
| Spangle-cheeked Tanager | Golden-browed Chlorophonia |
| Blue-hooded Euphonia | Spot-crowned Euphonia |
| Flame-colored Tanager | White-winged Tanager |
| Common Bush-Tanager | Sooty-capped Bush-Tanager |
| Buff-throated Saltator | Yellow-thighed Finch |
| Large-footed Finch | Yellow-throated Brush-Finch |
| Chestnut-capped Brush-Finch | Yellow-faced Grassquit |
| Rufous-collared Sparrow | Eastern Meadowlark |
| Yellow-bellied Siskin | Lesser Goldfinch |

**Fortuna**

    Construction of the Fortuna reservoir and trans-isthmian oil pipeline opened access to an interesting area of highland and foothill forest. The forest reserve is protected as a catchment zone for the reservoir, so good habitat is easy to reach.

    The highway through the area leaves the Pan-American 14 km east of David. It is paved and in good condition most of the 93 km to Chiriquí Grande on the Bocas coast. The nearest accommodations are in David and Chiriquí Grande, though there are a couple of restaurants about 15 km towards David from the dam. For information on the Caribbean slope and the Oleoducto Road, see page 137.

    With the conversion of the former Umbrellabird Road into the main highway, birding the Fortuna area mainly involves working along the paved road (little traffic) and a few side roads. The old highway, which now is interrupted by the lake, and a gated gravel road to the left a kilometer past the dam provide opportunities to leave the highway.

    At the pass there is a small checkpoint and a road off to the left. You can continue along this road for about 5 km, looking for trails. One that has been good in the past leaves the road to the left near some abandoned trucks about 3 km in. Birding along the road itself is good also.

## Birds of the Fortuna area

Little Tinamou
Am. Swallow-tailed Kite
Great Black-Hawk
Red-tailed Hawk
Ornate Hawk-Eagle
Bat Falcon
Crested Guan-r
Black-breasted Wood-Quail
Band-tailed Pigeon
Chiriquí Quail-Dove
Sulphur-winged Parakeet
Tropical Screech-Owl
Mottled Owl
Pauraque
White-collared Swift
Band-tailed Barbthroat
Little Hermit
Violet Sabrewing
Green Thorntail
White-tailed Emerald
White-bellied Mountain-gem
Magnificent Hummingbird
Resplendent Quetzal
Red-headed Barbet
Emerald Toucanet
Smoky-brown Woodpecker
Red-faced Spinetail
Ruddy Treerunner
Lineated Foliage-gleaner
Ruddy Foliage-gleaner
Tawny-throated Leaftosser
Strong-billed Woodcreeper-r

Fasciated Tiger-Heron-r
Barred Hawk
Short-tailed Hawk
Black Hawk-Eagle
Barred Forest-Falcon
Black Guan
Great Curassow-r
Spotted Sandpiper
Ruddy Pigeon
Crimson-fronted Parakeet
Blue-headed Parrot
Bare-shanked Screech-Owl
Black-and-White Owl
Chestnut-collared Swift
Vaux's Swift
Green Hermit
Green-fronted Lancebill
Green Violetear
Stripe-tailed Hummingbird
Variable Mountain-gem
Green-crowned Brilliant
Magenta-throated Woodstar
Orange-bellied Trogon
Prong-billed Barbet
Acorn Woodpecker
Golden-olive Woodpecker
Spotted Barbtail
Buffy Tuftedcheek
Spectacled Foliage-gleaner
Streak-breasted Treehunter
Olivaceous Woodcreeper
Spotted Woodcreeper

| | |
|---|---|
| Streak-headed Woodcreeper | Brown-billed Scythebill |
| Russet Antshrike | Slaty Antwren |
| Rufous-rumped Antwren-r | Immaculate Antbird |
| Rufous-breasted Antthrush | White-fronted Tyrannulet |
| Paltry Tyrannulet | Yellow-bellied Elaenia |
| Lesser Elaenia | Mountain Elaenia |
| Torrent Tyrannulet | Slaty-capped Flycatcher |
| Tufted Flycatcher | Dark Pewee |
| Scale-crested Pygmy-Tyrant | White-throated Spadebill |
| Bran-colored Flycatcher | Black Phoebe |
| Yellowish Flycatcher | Dusky-capped Flycatcher |
| Golden-bellied Flycatcher | White-winged Becard |
| Black-and-White Becard-r | Rufous Piha |
| Bare-necked Umbrellabird-r | Three-wattled Bellbird |
| Thrushlike Manakin | White-ruffed Manakin |
| White-crowned Manakin | Blue-and-White Swallow |
| S. Rough-winged Swallow | Azure-hooded Jay |
| Plain Wren | Ochraceous Wren |
| Gray-breasted Wood-Wren | S. Nightingale Wren |
| N. Am. Dipper | Black-faced Solitaire |
| Mountain Thrush | Pale-vented Thrush |
| Clay-colored Thrush | White-throated Thrush |
| Black-and-Yellow Silky-Flycatcher | Long-tailed Silky-Flycatcher |
| Yellow-throated Vireo | Brown-capped Vireo |
| Philadelphia Vireo | Lesser Greenlet |
| Golden-winged Warbler | Tennessee Warbler |
| Tropical Parula | Chestnut-sided Warbler |
| Blackburnian Warbler | Black-throated Green Warbler |
| Louisiana Waterthrush | MacGillivray's Warbler-r |
| Wilson's Warbler | Canada Warbler |
| Slate-throated Whitestart | Three-striped Warbler |
| Wrenthrush-r | Bananaquit |
| Silver-throated Tanager | Bay-headed Tanager |

Spangle-cheeked Tanager
Blue-hooded Euphonia
Red-crowned Ant-Tanager
Summer Tanager
White-winged Tanager
Sooty-capped Bush-Tanager
Buff-throated Saltator
Sooty-faced Finch-r
Slaty Finch-r
Rufous-collared Sparrow

Golden-browed Chlorophonia
Tawny-capped Euphonia
Hepatic Tanager
Flame-colored Tanager
Common Bush-Tanager
Streaked Saltator
Black-thighed Grosbeak
Yellow-throated Brush-Finch
Slaty Flower-piercer

## Chorcha Abajo

Across the Pan-American from the turn to Fortuna and Chiriquí Grande, a road goes southward into the ranch country typical of lowland Chiriquí. The habitat here is not too good, and were it not for the paucity of forest in the lowlands, would not really merit mention.

Nonetheless, in the patchy woodland remaining along the road and among the pastures, some species unlikely in the highlands can still be found. Of interest are Blue-throated Goldentail, Pale-eyed Pygmy-Tyrant, and Rose-throated Becard.

## Cerro Colorado

Cerro Colorado in eastern Chiriquí province is an area of considerable interest to birders, as it is the only easily-accessible area where two of Panama's endemic species can be found. The Glow-throated Hummingbird and the Yellow-green Finch, despite their very restricted ranges, are relatively easy to see here.

The road up to the mountain leaves the Pan-American Highway at the town of San Felix, 116 km west of Santiago and 74 km east of David. There are no accommodations closer than those cities, so start early to reach San Felix at first light. Otherwise it is not necessary to try to be on the mountain too early in the morning. There is a gas station and a restaurant (opens 6 a.m.) at the intersection.

Proceeding northward from San Felix, there is a bridge after 9 km with a gate and guardpost, where you will have to sign a register to enter. No permits are required. The road up the mountain is generally in good condition but there are a few rough and loose spots that might be difficult without 4WD. After 21 km there is a fork by some houses; continue to the right.

As the road climbs higher onto the mountain you will start to leave the deforested lowlands and reach a montane forest. Nearer the top look for any good habitat not chopped up by charcoal burners. Wind and mist can be a problem. One area that seems sheltered from the wind is the 500 meters or so of road before a small radio antenna.

A number of the highland species below are found here at a lower elevation than elsewhere in their ranges.

## Birds of Cerro Colorado

| | |
|---|---|
| Little Tinamou | Swallow-tailed Kite |
| Laughing Falcon | Scaled Pigeon |
| Band-tailed Pigeon | White-tipped Dove |
| Green Violetear | Variable Mountain-gem |
| Magnificent Hummingbird | Glow-throated Hummingbird |
| Orange-bellied Trogon | Prong-billed Barbet |
| Acorn Woodpecker | Spotted Barbtail |
| Ruddy Treerunner | Streak-headed Woodcreeper |
| Silvery-fronted Tapaculo | Mountain Elaenia |
| Blue-and-White Swallow | Plain Wren |
| Ochraceous Wren | Gray-breasted Wood-Wren |
| Black-faced Solitaire | Mountain Thrush |
| Rufous-browed Peppershrike | Brown-capped Vireo |
| Black-and-Yellow Silky-Flycatcher | Black-throated Green Warbler |
| Wilson's Warbler | Slate-throated Whitestart |
| Collared Whitestart | Rufous-capped Warbler |
| Black-cheeked Warbler | Spangle-cheeked Tanager |
| Hepatic Tanager | Flame-colored Tanager |
| Common Bush-Tanager | Sooty-capped Bush-Tanager |
| Yellow-green Finch | Chestnut-capped Brush-Finch |
| Rufous-collared Sparrow | Slaty Finch-r |
| Slaty Flower-piercer | Lesser Goldfinch |

## Las Lajas Marsh

Opposite the turn-off for San Felix, a paved road goes towards the town of Las Lajas and on to the beach. About 10 km from the highway, the road crosses a marsh that can be interesting in the northern winter. Shortly before the beach there is a dirt road to the left. Follow this to the back side of the marshy area.

# Bocas del Toro Province

The only portion of Bocas del Toro province readily accessible from the rest of Panama is the zone between Fortuna and the town of Chiriquí Grande. To reach the other areas of the province, one must take a ferry from Chiriquí Grande (about $40 each way for a car), fly from Panama City or David, or cross the border from Sixaola in Costa Rica.

The area has a number of specialties that reach the southern edge of their ranges here, though these are more easily found in Costa Rica or farther north. Some of these can be seen in the lowlands and foothills reached by the Oleoducto Road, an excellent birding area in any case.

## Oleoducto Road and Chiriquí Grande

The Oleoducto Road, so called because it generally follows the oil pipeline, provides a transect of lowland to foothill habitats. Even where the habitat is damaged by slash-and-burn farming and conversion to pasture, there is good birding. It is possible to drive along listening for the flocks of tanagers and other birds that are the main attraction. Twenty species of tanagers in a day is a real possibility. It is best to spend the night in one of the hotels in Chiriquí Grande and work the area from bottom to top.

## Birds of Oleoducto Road

| | |
|---|---|
| Little Tinamou | King Vulture |
| Swallow-tailed Kite | Black-shouldered Kite |
| Barred Hawk | White Hawk |
| Common Black-Hawk | Great Black-Hawk |
| Short-tailed Hawk | Black Hawk-Eagle |
| Bat Falcon | Gray-headed Chachalaca |
| White-throated Crake | Gray-breasted Crake |
| Gray-necked Wood-Rail | Northern Jacana |
| Pale-vented Pigeon | Scaled Pigeon |
| Short-billed Pigeon | Purplish-backed Quail-Dove-r |
| Crimson-fronted Parakeet | Sulphur-winged Parakeet |
| Red-fronted Parrotlet-r | Brown-hooded Parrot |

| | |
|---|---|
| Blue-headed Parrot | White-crowned Parrot |
| Red-lored Parrot | Mealy Parrot |
| Squirrel Cuckoo | Groove-billed Ani |
| Gray-rumped Swift | L. Swallow-tailed Swift |
| Bronzy Hermit | Band-tailed Barbthroat |
| Long-tailed Hermit | Little Hermit |
| White-necked Jacobin | Black-throated Mango |
| Violet-crowned Woodnymph | Blue-chested Hummingbird |
| Snowy-bellied Hummingbird | Rufous-tailed Hummingbird |
| Green-crowned Brilliant | Purple-crowned Fairy |
| Long-billed Starthroat | Violaceous Trogon |
| Slaty-tailed Trogon | Lattice-tailed Trogon-r |
| Rufous Motmot | Broad-billed Motmot |
| Green Kingfisher | Pied Puffbird |
| Lanceolated Monklet-r | Rufous-tailed Jacamar |
| Red-headed Barbet | Emerald Toucanet |
| Collared Aracari | Yellow-eared Toucanet |
| Keel-billed Toucan | Chestnut-mandibled Toucan |
| Olivaceous Piculet | Black-cheeked Woodpecker |
| Red-crowned Woodpecker | Smoky-brown Woodpecker |
| Golden-olive Woodpecker | Cinnamon Woodpecker |
| Lineated Woodpecker | Olivaceous Woodcreeper |
| Wedge-billed Woodcreeper | Barred Woodcreeper |
| Buff-throated Woodcreeper | Black-striped Woodcreeper-r |
| Spotted Woodcreeper | Streak-headed Woodcreeper |
| Fasciated Antshrike | Slaty Antshrike |
| Russet Antshrike | Checker-throated Antwren |
| Rufous-rumped Antwren-r | Chestnut-backed Antbird |
| Black-faced Antthrush | Rufous-breasted Antthrush |
| Black-headed Antthrush-r | Paltry Tyrannulet |
| Slaty-capped Flycatcher | Rufous-browed Tyrannulet |
| Scale-crested Pygmy-Tyrant | Common Tody-Flycatcher |
| Black-headed Tody-Flycatcher | Ruddy-tailed Flycatcher |

Black-tailed Flycatcher
Eastern Wood-Pewee
Long-tailed Tyrant
Gray-capped Flycatcher
Sulphur-bellied Flycatcher
Cinnamon Becard
Masked Tityra
Rufous Piha
Snowy Cotinga
Sharpbill-r
White-crowned Manakin
Mangrove Swallow
Brown Jay
Band-backed Wren
Bay Wren
Plain Wren
S. Nightingale-Wren
Wood Thrush
Pale-vented Thrush
Yellow-throated Vireo
Green Shrike-Vireo
Tennessee Warbler
Blackburnian Warbler
Black-and-White Warbler
Mourning Warbler
Slate-throated Whitestart
Plain-colored Tanager
Silver-throated Tanager
Bay-headed Tanager
Scarlet-thighed Dacnis
Shining Honeycreeper
White-vented Euphonia
Palm Tanager

Bran-colored Flycatcher
Tropical Pewee
Social Flycatcher
Streaked Flycatcher
Piratic Flycatcher
White-winged Becard
Black-crowned Tityra
Lovely Cotinga-r
Purple-throated Fruitcrow
White-ruffed Manakin
Gray-breasted Martin
Black-chested Jay
Azure-hooded Jay
Black-throated Wren-r
Stripe-throated Wren
Gray-breasted Wood-Wren
Swainson's Thrush
Black-headed N.-Thrush-r
Gray Catbird
Philadelphia Vireo
Golden-winged Warbler
Chestnut-sided Warbler
Bay-breasted Warbler
American Redstart
Olive-crowned Yellowthroat
Buff-rumped Warbler
Emerald Tanager
Speckled Tanager
Golden-hooded Tanager
Green Honeycreeper
Olive-backed Euphonia
Tawny-capped Euphonia
Blue-and-Gold Tanager

| | |
|---|---|
| Olive Tanager | Sulphur-rumped Tanager |
| Tawny-crested Tanager | White-lined Tanager |
| Red-throated Ant-Tanager | Hepatic Tanager |
| Summer Tanager | Scarlet-rumped Tanager |
| Crimson-collared Tanager | Crimson-backed Tanager |
| Dusky-faced Tanager | Common Bush-Tanager |
| Yellow-throated Bush-Tanager | Ashy-throated Bush-Tanager-r |
| Black-and-Yellow Tanager | Buff-throated Saltator |
| Slate-colored Grosbeak | Black-faced Grosbeak |
| Rose-breasted Grosbeak | Yellow-throated Brush-Finch |
| Orange-billed Sparrow | Black-striped Sparrow |
| Variable Seedeater | White-collared Seedeater |
| Lesser Seed-Finch | Giant Cowbird |
| Black-cowled Oriole | Orchard Oriole |
| Yellow-tailed Oriole | Yellow-billed Cacique |
| Scarlet-rumped Cacique | Montezuma Oropendola |

# Checklist

## Primary sources:

*Birds of Costa Rica Locational Checklist.*
Prepared by Costa Rica Expeditions. 1988.
*Birder's Field Checklist of the Birds of Panama.*
By Dodge Engleman and Horace Loftin. Russ' Natural History Books. 1992.
*An Annotated Checklist of the Birds of Monteverde and Peñas Blancas.* By Michael Fogden. 1993.
*A Guide to the Birds of Costa Rica.*
By Gary Stiles and Alexander Skutch. Cornell University Press. 1989.
*A Guide to the Birds of Panama.*
By Robert Ridgely and John Gwynne. Princeton University Press. 1989.

## The Areas:

Guanacaste: Generally the dry northwestern part of Costa Rica, including some areas around the Golfo de Nicoya in Puntarenas province.
Monteverde: The Cloud Forest Reserve and town of Monteverde, corresponding to Zones 2-4 in Michael Fogden's checklist.
Braulio Carrillo National Park: Generally the higher parts of the park, away from La Selva.
Finca La Selva: The OTS reserve and immediately adjacent areas.
Cerro de la Muerte. The massif above about 2000 meters, to include oak forest as well as the Cerro itself.
Corcovado: The Osa Peninsula.
Chiriquí: The highlands of Chiriquí province, including Fortuna.
Atlantic Canal Area: The Caribbean side of the old Canal Zone and vicinity including Pipeline Road.
Pacific Canal Area. The Pacific side of the Canal area and the Panama City area.
E. Foothills: Cerro Jefe/Azul principally.
Cana: The immediate area of Cana and Cerro Pirré.

## Abundance:

C: Common: Almost certain to be seen, usually in numbers, during two days of birding in the area.
U: Uncommon: 25-75% chance in two days in the area.
R: Rare: Normally present but less than 25% chance of locating the species in two days of birding.
X: Accidental or extirpated.

## Checklist of the Birds of Costa Rica and Panama

| | Costa Rica | Panama | Guanacaste | Monteverde | Braulio | La Selva | Cerro | Corcovado | Chiriquí | Atlantic | Pacific | E. Foothills | Cana |
|---|---|---|---|---|---|---|---|---|---|---|---|---|---|
| Great Tinamou | + | + | | | U | C | | C | | C | U | | U |
| Highland Tinamou | + | + | | U | R | | U | | R | | | | |
| Little Tinamou | + | + | | | C | C | | C | U | C | C | C | C |
| Thicket Tinamou | + | | C | | | | | | | | | | |
| Slaty-breasted Tinamou | + | | | | U | U | | | | | | | |
| Choco Tinamou | | + | | | | | | | | | | | R |
| Least Grebe | + | + | C | | | X | | U | R | C | C | | R |
| Pied-billed Grebe | + | + | C | | | | | U | R | C | C | | |
| Black Petrel | + | + | | | | | | | | | | | |
| Dark-rumped Petrel | + | + | | | | | | | | | | | |
| Black-capped Petrel | o | | | | | | | | | | | | |
| Wedge-tailed Shearwater | + | + | | | | | | | | | | | |
| Sooty Shearwater | + | + | | | | | | | | | | | |
| Short-tailed Shearwater | + | | | | | | | | | | | | |
| Pink-footed Shearwater | + | | | | | | | | | | | | |
| Cory's Shearwater | | o | | | | | | | | | | | |
| Audubon's Shearwater | + | + | | | | | | | | | | | |
| Wilson's Storm-Petrel | + | + | | | | | | | | | | | |
| Leach's Storm-Petrel | + | + | | | | | | | | | | | |
| Markham's Storm-Petrel | + | + | | | | | | | | | | | |
| Band-rumped Storm-Petrel | + | + | | | | | | | | | | | |
| Wedge-rumped Storm-Petrel | + | + | | | | | | | | | | | |
| Black Storm-Petrel | + | + | | | | | | | | | | | |
| Least Storm-Petrel | + | + | | | | | | | | | | | |
| White-faced Storm-Petrel | + | | | | | | | | | | | | |
| Red-billed Tropicbird | + | + | | | | | | | | | X | | |
| Masked Booby | + | + | | | | | | | | | | | |
| Blue-footed Booby | + | + | | | | | | | | | | | |
| Brown Booby | + | + | U | | | | | U | | | | | |
| Red-footed Booby | + | + | | | | | | | | | | | |
| Brown Pelican | + | + | C | | | | | U | | C | C | | |
| Am. White Pelican | o | o | X | | | | | | | | | | |
| Neotropic Cormorant | + | + | C | | | U | | U | R | C | C | | |
| Anhinga | + | + | U | | | R | | U | | | U | U | |
| Magnificent Frigatebird | + | + | C | X | | X | | C | | C | C | | |
| Great Frigatebird | + | | | | | | | | | | | | |

| | Costa Rica | Panama | Guanacaste | Monteverde | Braulio | La Selva | Cerro | Corcovado | Chiriquí | Atlantic | Pacific | E. Foothills | Cana |
|---|---|---|---|---|---|---|---|---|---|---|---|---|---|
| Pinnated Bittern | + | | R | | | | | | | | | | |
| American Bittern | o | o | X | | | | | | | X | | | |
| Least Bittern | + | + | U | | | X | | R | | R | R | | |
| Rufescent Tiger-Heron | + | + | | | R | R | | | | U | R | | R |
| Fasciated Tiger-Heron | + | + | | | | | R | | | R | | | R |
| Bare-throated Tiger-Heron | + | + | U | | | | | U | | R | | | |
| Great Blue Heron | + | + | U | | | R | | R | R | U | U | | |
| Cocoi Heron | | + | | | | | | | | R | | | |
| Great Egret | + | + | C | | | R | | U | U | C | C | R | |
| Snowy Egret | + | + | C | | | R | | R | R | C | C | R | |
| Little Blue Heron | + | + | C | | | C | | C | R | C | C | R | R |
| Tricolored Heron | + | + | U | | | R | | R | | C | C | | R |
| Reddish Egret | + | o | | | | | | R | | | | | |
| Cattle Egret | + | + | C | C | U | C | R | C | C | C | C | C | C |
| Green Heron | + | + | C | | | U | | C | U | C | C | | |
| Striated Heron | o | + | X | | | | | | | C | C | R | R |
| Agami Heron | + | + | | | | R | | R | | R | | | X |
| Capped Heron | | + | | | | | | | | R | | | |
| Black-crowned Night-Heron | + | + | U | | | R | | R | | U | U | | |
| Yellow-crowned Night-Heron | + | + | C | | | R | | C | X | C | C | | |
| Boat-billed Heron | + | + | C | | | R | | R | | U | U | | |
| White Ibis | + | + | C | | | | | U | | R | C | | |
| Glossy Ibis | + | + | U | | | | | | | R | R | | |
| Green Ibis | + | + | | | | R | | | | X | | | R |
| Roseate Spoonbill | + | + | C | | | | | U | | R | U | | |
| Jabiru | + | o | U | | | | | | | | | | |
| Wood Stork | + | + | C | X | | X | | R | | R | R | | |
| Fulvous Whistling-Duck | + | o | U | | | | | | | | | | |
| White-faced Whistling-Duck | o | o | R | | | | | | | X | | | |
| Black-bellied Whistling-Duck | + | + | C | | | R | | R | | R | R | | |
| Muscovy Duck | + | + | U | | | R | | R | | R | R | | |
| Comb Duck | | + | | | | | | | | | | | |
| Northern Pintail | + | o | U | | | | | | | X | | | |
| Blue-winged Teal | + | + | C | | | R | | R | U | U | U | | |
| Cinnamon Teal | o | o | R | | | | | | | X | X | | |
| Northern Shoveler | + | + | U | | | | | | | | X | | |
| American Wigeon | + | + | U | | | | | | | R | R | | |
| Ring-necked Duck | + | + | U | | | | | | X | R | R | | |

| | Costa Rica | Panama | Guanacaste | Monteverde | Braulio | La Selva | Cerro | Corcovado | Chiriquí | Atlantic | Pacific | E. Foothills | Cana |
|---|---|---|---|---|---|---|---|---|---|---|---|---|---|
| Lesser Scaup | + | + | R | | | | | | R | R | R | | |
| Greater Scaup | o | | X | | | | | | | | | | |
| Masked Duck | + | + | U | | | | | | R | R | R | R | |
| Black Vulture | + | + | C | C | C | C | U | C | C | C | C | C | C |
| Turkey Vulture | + | + | C | C | C | C | U | C | C | C | C | C | C |
| Lesser Yellow-headed Vulture | + | + | R | | | | | R | | | R | | |
| King Vulture | + | + | U | X | R | R | | U | | U | R | R | U |
| Osprey | + | + | C | | R | U | | C | U | C | C | | X |
| Gray-headed Kite | + | + | U | | | U | | U | R | U | U | R | R |
| Hook-billed Kite | + | + | R | R | | X | | R | | U | R | | |
| Am. Swallow-tailed Kite | + | + | X | C | C | U | R | U | C | U | U | C | C |
| Pearl Kite | | + | | | | | | | | R | R | R | R |
| White-tailed Kite | + | + | U | | | C | | C | U | C | C | R | |
| Snail Kite | + | o | U | | | | | R | | | X | | |
| Slender-billed Kite | | + | | | | | | | | | | | |
| Double-toothed Kite | + | + | R | R | U | U | | U | U | U | U | U | U |
| Mississippi Kite | + | + | | | | R | | | | R | R | R | R |
| Plumbeous Kite | + | + | U | | | R | | U | | R | R | R | U |
| Northern Harrier | + | + | R | | | R | | | | | X | | |
| Tiny Hawk | + | + | | | | R | | R | | R | | R | R |
| Sharp-shinned Hawk | + | + | R | R | U | | | | R | X | X | | |
| Bicolored Hawk | + | + | | R | R | R | | R | R | R | R | R | R |
| Cooper's Hawk | + | | | R | | | R | | | | | | |
| Crane Hawk | + | + | R | | | R | | R | | R | | R | R |
| Barred Hawk | + | + | | C | U | X | U | X | R | | | R | R |
| Plumbeous Hawk | | + | | | | | | | | R | | X | R |
| Semiplumbeous Hawk | + | + | | | | U | | | | U | R | | U |
| White Hawk | + | + | X | | U | U | | U | R | U | U | U | U |
| Common Black-Hawk | + | + | | | | R | | | | C | | | R |
| Mangrove Black-Hawk | + | + | C | R | | | | C | | | U | | |
| Great Black-Hawk | + | + | U | R | | U | | U | R | R | R | R | U |
| Savannah Hawk | o | + | | | | | | | | R | U | | |
| Harris' Hawk | + | o | U | | | | | | | | X | | |
| Black-collared Hawk | + | + | R | | | | | R | | | X | | |
| Solitary Eagle | + | + | | | R | X | | R | | | | | X |
| Gray Hawk | + | + | C | R | | X | | U | U | U | U | U | U |
| Roadside Hawk | + | + | C | R | | R | | U | U | R | R | R | U |
| Broad-winged Hawk | + | + | C | C | C | C | C | C | C | C | C | C | C |

145

| | Costa Rica | Panama | Guanacaste | Monteverde | Braulio | La Selva | Cerro | Corcovado | Chiriquí | Atlantic | Pacific | E. Foothills | Cana |
|---|---|---|---|---|---|---|---|---|---|---|---|---|---|
| Short-tailed Hawk | + | + | R | R | U | R | | R | R | U | U | U | R |
| Swainson's Hawk | + | + | R | | C | C | C | R | C | U | U | U | U |
| White-tailed Hawk | + | + | R | | | | | | R | | X | | |
| Zone-tailed Hawk | + | + | U | R | | | | R | R | R | U | U | R |
| Red-tailed Hawk | + | + | R | X | R | | U | | U | | R | | |
| Crested Eagle | o | o | | | | X | | | X | X | X | X | X |
| Harpy Eagle | o | + | | X | | X | | R | | X | X | X | X |
| Black-and-White Hawk-Eagle | + | + | | | | R | | R | R | R | | R | R |
| Black Hawk-Eagle | + | + | | R | U | U | | U | R | U | U | U | C |
| Ornate Hawk-Eagle | + | + | X | R | R | R | | R | X | R | R | R | R |
| Red-throated Caracara | + | + | | | | X | | R | | | X | X | U |
| Crested Caracara | + | + | C | | | | | U | | R | U | | |
| Yellow-headed Caracara | + | + | R | | | | | C | U | R | U | | |
| Laughing Falcon | + | + | C | | | U | | C | R | R | R | | U |
| Barred Forest-Falcon | + | + | | U | U | U | | U | U | U | U | U | U |
| Slaty-backed Forest-Falcon | + | + | | | | R | | | R | | | | R |
| Collared Forest-Falcon | + | + | U | R | R | R | | U | R | R | R | R | R |
| American Kestrel | + | + | U | R | | R | | R | U | R | R | R | X |
| Merlin | + | + | X | | | | | X | X | R | R | R | |
| Aplomado Falcon | o | + | X | | | | | | | X | | | |
| Bat Falcon | + | + | | | R | R | | U | U | R | R | U | U |
| Orange-breasted Falcon | o | o | | | | | | | | | | X | X |
| Peregrine Falcon | + | + | U | X | | R | | U | | R | U | X | |
| Plain Chachalaca | + | | U | | | | | | | | | | |
| Gray-headed Chachalaca | + | + | | R | U | | R | | C | C | | | C |
| Black Guan | + | + | | C | C | | C | R | | | | | |
| Crested Guan | + | + | U | | U | U | | U | X | R | X | | U |
| Great Curassow | + | + | R | | | R | | U | X | R | | R | R |
| Buffy-crowned Wood-Partridge | + | | | | | | R | | | | | | |
| Marbled Wood-Quail | + | + | | | | | | U | | R | R | R | U |
| Rufous-fronted Wood-Quail | + | + | | R | R | | | | | | | R | R |
| Black-breasted Wood-Quail | + | + | U | U | | | | R | | | | | |
| Tacarcuna Wood-Quail | | + | | | | | | | | | | | |
| Spotted Wood-Quail | + | + | | U | R | U | | R | | | | | |
| Tawny-faced Quail | + | + | | | | | | | | | R | X | R |
| Crested Bobwhite | + | + | C | | | | | U | | X | | | |

|  | Costa Rica | Panama | Guanacaste | Monteverde | Braulio | La Selva | Cerro | Corcovado | Chiriquí | Atlantic | Pacific | E. Foothills | Cana |
|---|---|---|---|---|---|---|---|---|---|---|---|---|---|
| White-throated Crake | + | + |  |  | U | C |  | C | U | U | U | U | R |
| Ruddy Crake | o |  | X |  |  |  |  |  |  |  |  |  |  |
| Gray-breasted Crake | + | o |  |  |  | R |  |  |  | X | X |  |  |
| Gray-necked Wood-Rail | + | + | U |  | R | U |  | U | U | U | U | U | U |
| Rufous-necked Wood-Rail | + | + | R |  |  |  |  |  |  |  | R |  |  |
| Uniform Crake | + | o |  |  |  | R |  | R |  | X |  |  |  |
| Sora | + | + | R | X |  |  |  |  | R | R | R |  |  |
| Yellow-breasted Crake | + | + | R |  |  |  |  |  |  | X | R |  |  |
| Black Rail | + | o | R |  |  |  |  | X |  |  | X |  |  |
| Colombian Crake |  | + |  |  |  |  |  |  |  | X | R |  |  |
| Spotted Rail | + | o | R |  |  |  |  |  |  | X | X |  |  |
| Paint-billed Crake | o | o |  |  |  |  |  |  |  |  | X |  |  |
| Purple Gallinule | + | + | U | X |  | X |  | U | R | U | U |  |  |
| Common Moorhen | + | + | C |  |  | X |  | R | R | U | U |  |  |
| American Coot | + | + | U |  |  |  |  |  | R | R | R |  |  |
| Sungrebe | + | + |  |  |  | U |  | U |  | R | R |  |  |
| Sunbittern | + | + |  | R | R | R |  |  |  | R | X | R |  |
| Limpkin | + | + | U |  |  |  |  | X | R | R |  |  |  |
| Double-striped Thick-knee | + |  | U |  |  |  |  |  |  |  |  |  |  |
| Southern Lapwing |  | + |  |  |  |  |  |  |  | X | R | R |  |
| Black-bellied Plover | + | + | C |  |  |  |  | U |  | C | C |  |  |
| American Golden-Plover | + | + | R |  |  |  |  |  |  | R | R |  |  |
| Collared Plover | + | + | R |  |  |  |  | U |  | U | U |  |  |
| Snowy Plover | + | o | R |  |  |  |  |  |  |  |  |  |  |
| Wilson's Plover | + | + | C |  |  |  |  | C |  | C | C |  |  |
| Semipalmated Plover | + | + | C |  |  |  |  | C |  | C | C |  |  |
| Killdeer | + | + | U |  |  | U |  | U | R | U | U |  |  |
| American Oystercatcher | + | + | R |  |  |  |  | R |  | X | R |  |  |
| Black-necked Stilt | + | + | C |  |  |  |  | R |  | U | U |  |  |
| American Avocet | o | o | X |  |  |  |  |  |  |  |  |  |  |
| Northern Jacana | + | + | C |  |  | C |  | C | C |  |  |  |  |
| Wattled Jacana | o | + |  |  |  |  |  |  |  | C | C |  | U |
| Greater Yellowlegs | + | + | U |  |  | X |  | U | R | C | C |  |  |
| Lesser Yellowlegs | + | + | C |  |  |  |  | U |  | C | C |  |  |
| Solitary Sandpiper | + | + | U | R | R | R | R | U | U | U | U |  |  |
| Willet | + | + | C |  |  |  |  | C |  | C | C |  |  |
| Wandering Tattler | + | o |  |  |  |  |  | U |  | X | X |  |  |

| | Costa Rica | Panama | Guanacaste | Monteverde | Braulio | La Selva | Cerro | Corcovado | Chiriquí | Atlantic | Pacific | E. Foothills | Cana |
|---|---|---|---|---|---|---|---|---|---|---|---|---|---|
| Spotted Sandpiper | + | + | C | | U | C | | C | C | C | C | R | U |
| Upland Sandpiper | + | + | R | | | | | | | R | R | | |
| Whimbrel | + | + | C | | | | | C | | C | C | | |
| Long-billed Curlew | + | o | R | | | | | X | | X | | | |
| Hudsonian Godwit | o | o | | | | | | | | X | | | |
| Marbled Godwit | + | + | U | | | | | R | | R | U | | |
| Ruddy Turnstone | + | + | U | | | | | U | | U | C | | |
| Surfbird | + | + | U | | | | | U | | | U | | |
| Red Knot | + | + | R | | | | | R | | R | U | | |
| Sanderling | + | + | C | | | | | C | | R | U | | |
| Semipalmated Sandpiper | + | + | C | | | | | C | | C | C | | |
| Western Sandpiper | + | + | C | | | | | C | | C | C | | |
| Least Sandpiper | + | + | C | | | | | C | | C | C | | |
| White-rumped Sandpiper | + | + | | | | | | | | R | R | X | |
| Baird's Sandpiper | + | + | | | | | | R | | R | R | | |
| Pectoral Sandpiper | + | + | R | | | | | R | | U | U | | |
| Dunlin | o | o | X | | | | | X | | | X | | |
| Curlew Sandpiper | o | | X | | | | | | | | | | |
| Stilt Sandpiper | + | + | R | | | | | R | | R | R | | |
| Buff-breasted Sandpiper | + | + | | | | | | | | R | R | | |
| Ruff | o | o | X | | | | | | | | X | | |
| Short-billed Dowitcher | + | + | C | | | | | U | | C | C | | |
| Long-billed Dowitcher | + | + | R | | | | | | | R | R | | |
| Common Snipe | + | + | U | | | | | | R | U | U | | |
| Wilson's Phalarope | + | + | U | | | | | | | R | R | | |
| Red-necked Phalarope | + | + | R | | | | | R | | | R | | |
| Red Phalarope | + | | | | | | | | | | | | |
| Pomarine Jaeger | + | + | R | | | | | R | | R | R | | |
| Parasitic Jaeger | + | + | R | | | | | | | R | R | | |
| Long-tailed Jaeger | o | | X | | | | | | | | | | |
| South Polar Skua | o | o | | | | | | | | | | | |
| Laughing Gull | + | + | C | | | | | C | | C | C | | |
| Franklin's Gull | + | + | U | | | | | U | | R | U | | |
| Bonaparte's Gull | o | o | | | | | | X | | X | | | |
| Ring-billed Gull | + | + | R | | | | | | | R | R | | |
| Herring Gull | + | + | R | | | | | | | R | R | | |
| Lesser Black-backed Gull | | o | | | | | | | | X | | | |

|  | Costa Rica | Panama | Guanacaste | Monteverde | Braulio | La Selva | Cerro | Corcovado | Chiriquí | Atlantic | Pacific | E. Foothills | Cana |
|---|---|---|---|---|---|---|---|---|---|---|---|---|---|
| Sabine's Gull | + | + | R |  |  |  |  |  |  | X | R |  |  |
| Gray Gull | o | o |  |  |  |  |  |  |  |  | X |  |  |
| Band-tailed Gull |  | o |  |  |  |  |  |  |  |  | X |  |  |
| Gull-billed Tern | + | + | R |  |  |  |  | R |  | R | U |  |  |
| Caspian Tern | + | + | R |  |  |  |  |  |  | R | R |  |  |
| Royal Tern | + | + | C |  |  |  |  | U |  | C | C |  |  |
| Elegant Tern | + | + | U |  |  |  |  | R |  |  | R |  |  |
| Common Tern | + | + | U |  |  |  |  | U |  | U | C |  |  |
| Forster's Tern | + | + | R |  |  |  |  |  |  | R | R |  |  |
| Least Tern | + | + | U |  |  |  |  | U |  | U | U |  |  |
| Yellow-billed Tern |  | o |  |  |  |  |  |  |  |  | X |  |  |
| Bridled Tern | + | + | U |  |  |  |  | U |  |  | R |  |  |
| Sooty Tern | + | + |  |  | X |  |  | R |  | X | X |  |  |
| Large-billed Tern |  | o |  |  |  |  |  |  |  | X | X |  |  |
| Black Tern | + | + | C |  |  |  |  | C |  | U | U |  |  |
| Brown Noddy | + | + | X |  |  |  |  | R |  |  | R |  |  |
| Black Noddy | + |  |  |  |  |  |  |  |  |  |  |  |  |
| White Tern | + | o |  |  |  |  |  |  |  |  | X |  |  |
| Black Skimmer | + | + | U |  |  |  |  | U |  | R | R |  |  |
| Rock Dove | + | + |  |  |  |  |  |  | R | C | C |  |  |
| Pale-vented Pigeon | + | + |  |  |  | R |  | C | U | C | C | C | C |
| Scaled Pigeon | + | + |  |  |  | U |  |  | U | U | U | U | C |
| White-crowned Pigeon | o | + |  |  |  |  |  |  |  |  |  |  |  |
| Red-billed Pigeon | + |  | C | U |  | R |  | R |  |  |  |  |  |
| Band-tailed Pigeon | + | + |  | C | U |  | C |  | U |  |  |  |  |
| Ruddy Pigeon | + | + |  | C | C |  | U |  | U |  |  |  | U |
| Short-billed Pigeon | + | + |  | R | C | C |  | C | U | C | C | U | U |
| Dusky Pigeon |  | o |  |  |  |  |  |  |  |  |  |  | X |
| White-winged Dove | + | + | C |  |  |  |  |  |  |  |  |  |  |
| Mourning Dove | + | + | U |  |  | X |  | R | U |  | X |  |  |
| Inca Dove | + |  | C |  |  |  |  |  |  |  |  |  |  |
| Eared Dove |  | o |  |  |  |  |  |  |  |  | X |  |  |
| Common Ground-Dove | + | + | C |  |  |  |  |  |  |  |  |  |  |
| Plain-breasted Ground-Dove | + | + | R |  |  |  |  |  |  | R | U |  |  |
| Ruddy Ground-Dove | + | + | R |  |  | C |  | C | U | C | C |  |  |
| Blue Ground-Dove | + | + | R |  | R | C |  | C | U | U | U |  | C |
| Maroon-chested Ground-Dove | + | + |  |  |  |  | R |  | R |  |  |  |  |

| | Costa Rica | Panama | Guanacaste | Monteverde | Braulio | La Selva | Cerro | Corcovado | Chiriquí | Atlantic | Pacific | E. Foothills | Cana |
|---|---|---|---|---|---|---|---|---|---|---|---|---|---|
| White-tipped Dove | + | + | C | C | | X | | C | C | C | C | C | C |
| Gray-headed Dove | + | + | R | | | | | | | | | | |
| Gray-chested Dove | + | + | | | C | C | | C | U | C | C | | U |
| Olive-backed Quail-Dove | + | + | | | R | U | | | | R | | | |
| Chiriquí Quail-Dove | + | + | | U | R | | U | | U | | | | |
| Purplish-backed Quail-Dove | + | + | | | R | | | | | | | R | |
| Buff-fronted Quail-Dove | + | + | | U | R | | U | | U | | | | |
| Russet-crowned Quail-Dove | | + | | | | | | | | | | | U |
| Ruddy Quail-Dove | + | + | | | R | R | | U | R | R | R | R | U |
| Violaceous Quail-Dove | + | + | | | | | | | | R | | R | R |
| Painted Parakeet | | + | | | | | | | | | | | |
| Sulphur-winged Parakeet | + | + | | U | | | U | | C | | | | |
| Crimson-fronted Parakeet | + | + | | R | C | C | | C | U | | | | |
| Olive-throated Parakeet | + | + | | | | C | | | | | | | |
| Orange-fronted Parakeet | + | | C | R | | | | | | | | | |
| Brown-throated Parakeet | | + | | | | | | | | R | | | |
| Great Green Macaw | + | + | | | R | R | | | | X | | X | C |
| Scarlet Macaw | + | + | R | | | | | C | | | | | |
| Blue-and-Yellow Macaw | | + | | | | | | | | X | X | X | C |
| Red-and-Green Macaw | | + | | | | | | | | X | | X | C |
| Chestnut-fronted Macaw | | + | | | | | | | | | | X | C |
| Barred Parakeet | + | + | | | R | U | | U | U | | | | |
| Spectacled Parrotlet | | + | | | | | | | | | | R | |
| Orange-chinned Parakeet | + | + | C | | U | U | | C | C | C | C | C | C |
| Red-fronted Parrotlet | + | + | | R | R | | R | | R | | | | |
| Blue-fronted Parrotlet | | + | | | | | | | | | | R | R |
| Saffron-headed Parrot | | + | | | | | | | | | | | R |
| Brown-hooded Parrot | + | + | | C | C | C | | C | R | U | U | | C |
| Blue-headed Parrot | + | + | | | | | | R | U | C | C | C | C |
| White-crowned Parrot | + | + | | | C | C | | C | U | | | | |
| White-fronted Amazon | + | | C | U | | | | | | | | | |
| Red-lored Amazon | + | + | | R | U | C | | C | | C | C | U | R |
| Mealy Amazon | + | + | | | | C | | C | R | C | C | U | C |
| Yellow-naped Amazon | + | | U | | | | | | | | | | |
| Yellow-crowned Amazon | | + | | | | | | | | | R | | |
| Black-billed Cuckoo | + | + | R | R | | R | | R | R | R | R | | |
| Yellow-billed Cuckoo | + | + | R | R | R | R | | R | R | R | R | | U |

| | Costa Rica | Panama | Guanacaste | Monteverde | Braulio | La Selva | Cerro | Corcovado | Chiriquí | Atlantic | Pacific | E. Foothills | Cana |
|---|---|---|---|---|---|---|---|---|---|---|---|---|---|
| Dwarf Cuckoo | o | | | | | | | | | | X | | |
| Dark-billed Cuckoo | o | | | | | | | | | | X | | |
| Gray-capped Cuckoo | o | | | | | | | | | | X | | X |
| Mangrove Cuckoo | + | + | U | | | | | R | | R | R | | |
| Cocos Cuckoo | + | | | | | | | | | | | | |
| Squirrel Cuckoo | + | + | C | C | C | C | | C | U | C | C | R | C |
| Little Cuckoo | | + | | | | | | | | R | R | | U |
| Striped Cuckoo | + | + | | | | U | | U | R | R | U | R | U |
| Pheasant Cuckoo | + | + | R | | | | | | | R | R | | U |
| Lesser Ground-Cuckoo | + | | | U | R | | | | | | | | |
| Rufous-vented Ground-Cuckoo | + | + | | | | | R | | | R | | | R |
| Greater Ani | | + | | | | | | | | C | U | | C |
| Smooth-billed Ani | + | + | R | | | | | C | U | C | C | C | C |
| Groove-billed Ani | + | + | C | C | U | C | R | | | R | R | | |
| Barn Owl | + | + | U | R | | R | | R | R | R | R | | |
| Pacific Screech-Owl | + | | C | | | | | | | | | | |
| Vermiculated Screech-Owl | + | + | | | R | U | | | R | R | R | R | R |
| Tropical Screech-Owl | + | + | R | | | | | | R | R | U | U | |
| Bare-shanked Screech-Owl | + | + | | U | R | | U | | R | | | | R |
| Crested Owl | + | + | R | X | | R | | R | | R | R | | |
| Spectacled Owl | + | + | U | U | U | U | | U | U | U | U | | U |
| Great Horned Owl | + | o | X | | | | | | | | | | |
| Andean Pygmy-Owl | + | + | | | R | | U | | R | | | | |
| Least Pygmy-Owl | + | + | | | R | | | | | R | | R | U |
| Ferruginous Pygmy-Owl | + | + | U | | | | | | | X | | | |
| Burrowing Owl | o | o | | | | | | | | | | | |
| Mottled Owl | + | + | R | U | U | U | | U | U | U | U | U | U |
| Black-and-White Owl | + | + | R | | R | | | R | U | R | R | | |
| Striped Owl | + | + | R | | | | | | R | R | R | | |
| Unspotted Saw-whet Owl | + | o | | | | R | X | | | | | | |
| Short-tailed Nighthawk | + | + | | | | U | U | | R | X | | U | |
| Lesser Nighthawk | + | + | C | | | | U | R | U | C | R | | |
| Common Nighthawk | + | + | U | | R | | R | R | U | U | U | | |
| Pauraque | + | + | C | C | C | C | C | C | C | C | C | C | |
| Chuck-wills-widow | + | + | X | X | R | | R | R | R | R | | | |
| Rufous Nightjar | + | + | | | | | | | R | U | U | R | |

| | Costa Rica | Panama | Guanacaste | Monteverde | Braulio | La Selva | Cerro | Corcovado | Chiriquí | Atlantic | Pacific | E. Foothills | Cana |
|---|---|---|---|---|---|---|---|---|---|---|---|---|---|
| Whip-poor-will | + | o | R | | | | | X | | | | | |
| Dusky Nightjar | + | + | | R | | | R | | R | | | | |
| Ocellated Poorwill | + | | | | | | | | | | | | |
| White-tailed Nightjar | + | + | R | | | | | R | | | U | | |
| Great Potoo | + | + | | | | U | | R | | U | | | R |
| Common Potoo | + | | R | | | | | | | | | | |
| Gray Potoo | + | + | | | | U | | R | R | C | C | U | U |
| Oilbird | o | o | | | | | X | | | | X | | |
| Black Swift | + | o | | R | U | R | U | U | | | | | X |
| White-chinned Swift | + | o | | | | | | | | | | | |
| Spot-fronted Swift | + | | | | | | | | | | | | |
| Chestnut-collared Swift | + | + | | R | U | | U | | U | | R | R | |
| White-collared Swift | + | + | U | C | C | C | C | C | C | U | R | C | C |
| Chimney Swift | + | + | | X | | R | | | | | U | U | |
| Vaux's Swift | + | + | U | C | C | | C | | C | R | R | | |
| Chapman's Swift | | o | | | | | | | | X | | | X |
| Short-tailed Swift | | + | | | | | | | | U | C | | |
| Ashy-tailed Swift | | o | | | | | | | | | X | | |
| Band-rumped Swift | + | + | | | | | | C | R | C | U | C | C |
| Gray-rumped Swift | + | + | | | U | C | | | | | | | |
| Lesser Swallow-tailed Swift | + | + | | R | U | U | | C | U | C | C | | C |
| Great Swallow-tailed Swift | o | | | | | X | | | | | | | |
| Bronzy Hermit | + | + | | | | C | | C | | | | | |
| Rufous-breasted Hermit | | + | | | | | | | | C | U | U | C |
| Band-tailed Barbthroat | + | + | | | U | C | | C | R | U | R | U | U |
| Green Hermit | + | + | | C | C | R | R | | C | X | | C | C |
| Long-tailed Hermit | + | + | R | | C | C | | C | U | C | C | C | C |
| Pale-bellied Hermit | | + | | | | | | | | | R | | |
| Little Hermit | + | + | R | R | U | C | | C | U | C | C | C | C |
| White-tipped Sicklebill | + | + | | R | U | | | R | | X | | R | U |
| Tooth-billed Hummingbird | | + | | | | | | | | | | | R |
| Green-fronted Lancebill | + | + | | R | U | | R | | R | | | | R |
| Scaly-breasted Hummingbird | + | + | R | | | R | | C | | C | C | | |
| Violet Sabrewing | + | + | | C | C | | R | | U | | | | |
| White-necked Jacobin | + | + | | X | U | C | | C | U | C | C | U | U |
| Brown Violetear | + | + | | R | R | X | | R | R | | | | U |
| Green Violetear | + | + | | C | C | | C | | C | | | | |
| Green-breasted Mango | + | + | U | | | R | | U | | X | | | |

|  | Costa Rica | Panama | Guanacaste | Monteverde | Braulio | La Selva | Cerro | Corcovado | Chiriqui | Atlantic | Pacific | E. Foothills | Cana |
|---|---|---|---|---|---|---|---|---|---|---|---|---|---|
| Black-throated Mango | + |  |  |  |  |  |  |  |  | C | C |  | U |
| Violet-headed Hummingbird | + | + |  |  | C | C |  | U | R | R |  | C | R |
| Rufous-crested Coquette | o | + |  |  |  |  |  | R | R | R | R | R |  |
| Black-crested Coquette | + |  |  |  | R | R |  |  |  |  |  |  |  |
| White-crested Coquette | + | + | X |  |  |  |  | U | R |  |  |  |  |
| Green Thorntail | + | + |  | X | U | R |  |  | R |  |  | R | R |
| Fork-tailed Emerald | + |  | C | C |  |  |  |  |  |  |  |  |  |
| Garden Emerald | + | + |  |  |  |  |  | U | U | R | U | U |  |
| Violet-crowned Woodnymph | + | + |  |  | U | C |  | C | U | C | U | C |  |
| Green-crowned Woodnymph |  | + |  |  |  |  |  |  |  |  |  |  | C |
| Fiery-throated Hummingbird | + | + |  |  | C |  | C |  | C |  |  |  |  |
| Violet-bellied Hummingbird |  | + |  |  |  |  |  |  |  | C | U | U |  |
| Sapphire-thrtd Hummingbird |  | + |  |  |  |  |  |  |  | C | C |  |  |
| Blue-headed Sapphire |  | + |  |  |  |  |  |  |  |  |  |  |  |
| Blue-throated Goldentail | + | + | U | R | U | R |  | C | R | R |  |  | C |
| Violet-capped Hummingbird |  | + |  |  |  |  |  |  |  |  |  | R |  |
| Rufous-cheeked Hummingbird |  | + |  |  |  |  |  |  |  |  |  |  | U |
| White-bellied Emerald | o |  |  |  |  |  | X |  |  |  |  |  |  |
| Blue-chested Hummingbird | + | + |  |  | U | C |  |  |  | C | U | R | U |
| Charming Hummingbird | + | + |  |  |  |  |  | C |  |  |  |  |  |
| Mangrove Hummingbird | + |  | R |  |  |  |  | U |  |  |  |  |  |
| Blue-tailed Hummingbird | o |  |  |  |  | X |  |  |  |  |  |  |  |
| Steely-vented Hummingbird | + |  | C | C |  |  |  |  |  |  |  |  |  |
| Snowy-bellied Hummingbird | + | + |  |  |  |  |  |  | U |  | C |  |  |
| Rufous-tailed Hummingbird | + | + | U | C | U | C |  | C | C | C | C | C | R |
| Cinnamon Hummingbird | + |  | C |  |  | X |  |  |  |  |  |  |  |
| Stripe-tailed Hummingbird | + | + |  | C | R |  | R |  | U |  |  |  |  |
| Black-bellied Hummingbird | + | + |  |  | U |  |  |  | R |  |  |  |  |
| White-tailed Emerald | + | + |  |  |  |  |  |  | U |  |  |  |  |
| Coppery-headed Emerald | + |  |  | C | U |  |  |  |  |  |  |  |  |
| Snowcap | + | + |  |  | U | U |  |  |  |  |  |  |  |
| White-vented Plumeleteer |  | + |  |  |  |  |  |  |  | R | U |  | C |
| Bronze-tailed Plumeleteer | + | + |  |  |  | C |  |  |  | R | R | U | R |
| White-bellied Mountain-gem | + | + | U | C |  |  |  | R |  |  |  |  |  |
| Variable Mountain-gem | + | + |  | C | C |  | C | C |  |  |  |  |  |
| Green-crowned Brilliant | + | + |  | C | C |  | R | U |  |  |  | R | R |
| Magnificent Hummingbird | + | + |  |  |  | R | U | U |  |  |  |  |  |
| Purple-crowned Fairy | + | + |  |  | U | U |  | U | R | C | C | U | C |

|  | Costa Rica | Panama | Guanacaste | Monteverde | Braulio | La Selva | Cerro | Corcovado | Chiriquí | Atlantic | Pacific | E. Foothills | Cana |
|---|---|---|---|---|---|---|---|---|---|---|---|---|---|
| Greenish Puffleg |  | + |  |  |  |  |  |  |  |  |  |  | U |
| Long-billed Starthroat | + | + |  |  |  | R |  | U | U | R | R | R | R |
| Plain-capped Starthroat | + |  | U | R |  |  |  |  |  |  |  |  |  |
| Magenta-throated Woodstar | + | + |  |  | C |  |  |  | R |  |  |  |  |
| Ruby-throated Hummingbird | + | + | C | R |  | R |  | C |  |  |  |  |  |
| Purple-throated Woodstar |  | o |  |  |  |  |  |  |  |  |  |  | X |
| Volcano Hummingbird | + | + |  |  |  |  | C |  | C |  |  |  |  |
| Glow-throated Hummingbird |  | + |  |  |  |  |  |  | R |  |  |  |  |
| Scintillant Hummingbird | + | + |  | U | U |  |  |  | C |  |  |  |  |
| Black-headed Trogon | + |  | C |  |  |  |  |  |  |  |  |  |  |
| White-tailed Trogon |  | + |  |  |  |  |  |  |  | C | X | R | C |
| Baird's Trogon | + | + |  |  |  |  |  | C |  |  |  |  |  |
| Violaceous Trogon | + | + | U |  | U | C |  | U | U | C | C | C | C |
| Elegant Trogon | + |  | C |  |  |  |  |  |  |  |  |  |  |
| Collared Trogon | + | + |  |  |  | C | U |  | C |  |  |  | C |
| Orange-bellied Trogon | + | + |  | C | R |  |  |  | U |  |  |  |  |
| Black-throated Trogon | + | + | X |  | C | C |  | C | U | C | C | R | C |
| Black-tailed Trogon |  | + |  |  |  |  |  |  |  | R | R |  | U |
| Slaty-tailed Trogon | + | + |  |  | C | C |  | C | R | C | C | R | R |
| Lattice-tailed Trogon | + | + |  |  | U | R |  |  | R |  |  |  |  |
| Golden-headed Quetzal |  | + |  |  |  |  |  |  |  |  |  |  | U |
| Resplendent Quetzal | + | + |  | C | U |  | C |  | C |  |  |  |  |
| Tody Motmot | + | + |  |  |  |  |  |  |  |  |  | R | U |
| Blue-crowned Motmot | + | + | U | C |  |  |  | C | U | U | C |  |  |
| Rufous Motmot | + | + |  |  | U | C |  |  |  | C | U | U | U |
| Keel-billed Motmot | + |  |  |  |  |  |  |  |  |  |  |  |  |
| Broad-billed Motmot | + | + |  | X | C | C |  |  |  | U |  | U | U |
| Turquoise-browed Motmot | + |  | C |  |  |  |  |  |  |  |  |  |  |
| Ringed Kingfisher | + | + | U |  |  | U |  | C | R | C | C |  |  |
| Belted Kingfisher | + | + | U |  |  | R |  | U | R | U | U |  |  |
| Green Kingfisher | + | + | U | R | C | C |  | C | U | C | C |  | U |
| Amazon Kingfisher | + | + | R |  | U | U |  | U |  | R | U |  |  |
| Green-and-Rufous Kingfisher | + | + |  |  |  | R |  |  |  | R | R |  |  |
| Am. Pygmy Kingfisher | + | + | U |  |  | U |  | U |  | U | U |  | U |
| Barred Puffbird |  | + |  |  |  |  |  |  |  |  |  |  | C |
| White-necked Puffbird | + | + | U |  | R | R |  | U |  | U | U |  | R |
| Black-breasted Puffbird |  | + |  |  |  |  |  |  |  | U | U | R | U |
| Pied Puffbird | + | + |  |  |  | R |  |  |  | U |  |  | C |

| | Costa Rica | Panama | Guanacaste | Monteverde | Braulio | La Selva | Cerro | Corcovado | Chiriquí | Atlantic | Pacific | E. Foothills | Cana |
|---|---|---|---|---|---|---|---|---|---|---|---|---|---|
| White-whiskered Puffbird | + | + | | | U | U | | C | X | U | U | R | U |
| Lanceolated Monklet | + | o | | | R | | | | | | | | X |
| Gray-cheeked Nunlet | | + | | | | | | | R | | | | C |
| White-fronted Nunbird | + | + | | | U | C | | | X | | R | | C |
| Dusky-backed Jacamar | | + | | | | | | | | | | | U |
| Rufous-tailed Jacamar | + | + | | X | U | C | | C | R | | | R | |
| Great Jacamar | + | + | | | | R | | | R | R | R | | U |
| Spot-crowned Barbet | | + | | | | | | | U | R | R | | C |
| Red-headed Barbet | + | + | | R | C | | | | U | | | | U |
| Prong-billed Barbet | + | + | | C | C | | C | | U | | | | |
| Emerald Toucanet | + | + | | C | C | X | C | | C | | | R | C |
| Collared Aracari | + | + | R | X | C | C | | | | C | C | U | C |
| Fiery-billed Aracari | + | + | | | | | | C | U | | | | |
| Yellow-eared Toucanet | + | + | | | U | R | | | R | | | U | C |
| Keel-billed Toucan | + | + | R | C | R | C | | | | C | C | C | C |
| Chestnut-mandibled Toucan | + | + | | | | C | | C | U | C | C | C | C |
| Olivaceous Piculet | + | + | | | | | | R | U | R | R | | C |
| Acorn Woodpecker | + | + | | | | | C | | C | | | | |
| Golden-naped Woodpecker | + | + | | | | | | C | X | | | | |
| Black-cheeked Woodpecker | + | + | | | U | C | | | | C | C | C | C |
| Red-crowned Woodpecker | + | + | | | | | | C | C | C | C | | |
| Hoffman's Woodpecker | + | | C | C | | | | | | | | | |
| Yellow-bellied Sapsucker | + | + | R | R | | | | R | | R | R | R | |
| Hairy Woodpecker | + | + | | | U | R | U | | U | | | | |
| Smoky-brown Woodpecker | + | + | | | U | U | | | R | | | | |
| Red-rumped Woodpecker | + | + | | | | | | R | | | | | C |
| Rufous-winged Woodpecker | + | + | | | | U | U | U | | | | | |
| Stripe-cheeked Woodpecker | | + | | | | | | | | | | R | U |
| Golden-green Woodpecker | | + | | | | | | | | | | | R |
| Golden-olive Woodpecker | + | + | | | C | C | | | U | | | | |
| Spot-breasted Woodpecker | | + | | | | | | | | R | | | |
| Cinnamon Woodpecker | + | + | | | U | U | | | | U | U | R | U |
| Chestnut-collared Woodpecker | + | + | | | U | U | | | | | | | |
| Lineated Woodpecker | + | + | U | R | U | U | | U | U | C | C | C | R |
| Crimson-bellied Woodpecker | | + | | | | | | | | R | | R | U |
| Crimson-crested Woodpecker | | + | | | | | | | | C | C | C | C |
| Pale-billed Woodpecker | + | + | U | R | U | C | | C | R | | | | |

| | Costa Rica | Panama | Guanacaste | Monteverde | Braulio | La Selva | Cerro | Corcovado | Chiriquí | Atlantic | Pacific | E. Foothills | Cana |
|---|---|---|---|---|---|---|---|---|---|---|---|---|---|
| Pale-breasted Spinetail | + | + | | | | | | | U | | U | | |
| Slaty Spinetail | + | + | | | R | C | | C | U | R | | | C |
| Red-faced Spinetail | + | + | | C | C | | R | | U | | | | U |
| Coiba Spinetail | | + | | | | | | | | | | | |
| Double-banded Graytail | | + | | | | | | | | | | | R |
| Spotted Barbtail | + | + | | C | C | | U | | C | | | | U |
| Beautiful Treerunner | | + | | | | | | | | | | | R |
| Ruddy Treerunner | + | + | | U | C | | C | | C | | | | |
| Buffy Tuftedcheek | + | + | | R | U | | U | | U | | | | |
| Striped Woodhaunter | + | + | | | U | U | | U | | | | R | |
| Lineated Foliage-gleaner | + | + | | U | U | | | | U | | | | U |
| Spectacled Foliage-gleaner | + | + | | | U | | | | U | | | | |
| Slaty-winged Foliage-gleaner | | + | | | | | | | | R | | R | U |
| Buff-fronted Foliage-gleaner | + | + | | | U | | R | | R | | | | |
| Buff-throated Foliage-gleaner | + | + | | | U | C | | U | R | U | | C | U |
| Ruddy Foliage-gleaner | + | + | | | | | | | U | | | R | |
| Streak-breasted Treehunter | + | + | | C | U | | U | | U | | | | |
| Plain Xenops | + | + | R | | C | U | | C | U | C | C | C | C |
| Streaked Xenops | + | + | | | R | | U | | R | | | | R |
| Tawny-throated Leaftosser | + | + | | R | R | | | | R | R | | | R |
| Gray-throated Leaftosser | + | + | | R | U | | | | R | | | | |
| Scaly-throated Leaftosser | + | + | | | R | U | | R | R | R | | | R |
| Sharp-tailed Streamcreeper | | + | | | | | | | | | | | R |
| Plain-brown Woodcreeper | + | + | | | R | U | | | | C | | U | C |
| Tawny-winged Woodcreeper | + | + | | | | | | C | R | | | | |
| Ruddy Woodcreeper | + | + | U | R | | | | | R | R | R | U | |
| Olivaceous Woodcreeper | + | + | U | C | U | | | | R | U | U | R | U |
| Long-tailed Woodcreeper | + | + | | | | X | | U | | R | R | R | U |
| Wedge-billed Woodcreeper | + | + | | X | C | C | | C | R | U | U | U | C |
| Strong-billed Woodcreeper | + | + | | R | R | | | | R | | | | |
| Barred Woodcreeper | + | + | R | R | U | C | | C | | R | | U | U |
| Black-banded Woodcreeper | + | + | | R | R | | | | R | | | | |
| Straight-billed Woodcreeper | | + | | | | | | | | U | U | | |
| Buff-throated Woodcreeper | + | + | R | | U | C | | C | | C | C | C | C |
| Ivory-billed Woodcreeper | + | | U | | | | | | | | | | |
| Black-striped Woodcreeper | + | + | | | U | U | | U | | U | R | U | |
| Spotted Woodcreeper | + | + | | C | C | U | | R | R | | | C | U |

| | Costa Rica | Panama | Guanacaste | Monteverde | Braulio | La Selva | Cerro | Corcovado | Chiriqui | Atlantic | Pacific | E. Foothills | Cana |
|---|---|---|---|---|---|---|---|---|---|---|---|---|---|
| Streak-headed Woodcreeper | + | + | U | C | U | C | | C | U | R | R | U | U |
| Spot-crowned Woodcreeper | + | + | | R | U | | C | | C | | | | |
| Red-billed Scythebill | | + | | | | | | | | R | X | | U |
| Brown-billed Scythebill | + | + | | R | U | | | R | R | | | R | U |
| Fasciated Antshrike | + | + | | | R | U | | | | U | U | C | C |
| Great Antshrike | + | + | | | | C | | C | R | U | U | | C |
| Barred Antshrike | + | + | U | | | R | | U | | C | C | U | |
| Black Antshrike | | + | | | | | | | | | | | U |
| Black-hooded Antshrike | + | + | | | | | | C | | | | | |
| Slaty Antshrike | + | + | | | | C | | | | C | C | C | C |
| Speckled Antshrike | | + | | | | | | | | | | R | |
| Russet Antshrike | + | + | | | R | U | | U | R | R | | R | U |
| Plain Antvireo | + | + | | C | U | X | R | U | R | | | U | U |
| Streak-crowned Antvireo | + | | | | U | U | | | | | | | |
| Spot-crowned Antvireo | + | + | | | | | | | | U | | U | |
| Pygmy Antwren | | + | | | | | | | | U | R | | U |
| Streaked Antwren | | + | | | | | | | | U | U | | C |
| Checker-throated Antwren | + | + | | | U | C | | | | C | U | | C |
| White-flanked Antwren | + | + | | | | C | | | | C | U | | C |
| Slaty Antwren | + | + | | | C | C | | R | C | U | | U | C |
| Rufous-winged Antwren | | + | | | | | | | | | | | C |
| Dot-winged Antwren | + | + | | | U | C | | C | | C | U | | C |
| White-fringed Antwren | | + | | | | | | | | | | | |
| Rufous-rumped Antwren | + | + | | | R | | | | R | | | | U |
| Dusky Antbird | + | + | | | U | C | | C | | C | C | | C |
| Jet Antbird | | + | | | | | | | | U | U | | U |
| Bare-crowned Antbird | + | + | | | | R | | U | | U | R | | U |
| White-bellied Antbird | | + | | | | | | | | R | U | | |
| Chestnut-backed Antbird | + | + | R | | U | C | | C | | C | U | | C |
| Dull-mantled Antbird | + | + | | | | R | | | | R | | R | |
| Immaculate Antbird | + | + | | R | U | X | | | | | | | R |
| Spotted Antbird | + | + | | | R | U | | | | C | U | U | R |
| Wing-banded Antbird | | + | | | | | | | | R | | | U |
| Bicolored Antbird | + | + | | | U | U | | U | | U | U | U | C |
| Ocellated Antbird | + | + | | | | U | | | | R | R | R | U |
| Black-faced Antthrush | + | + | | R | R | C | | C | R | C | C | C | C |
| Black-headed Antthrush | + | + | | | R | | | | R | | | R | |
| Rufous-breasted Antthrush | + | + | | R | R | | | | R | | | | U |

| | Costa Rica | Panama | Guanacaste | Monteverde | Braulio | La Selva | Cerro | Corcovado | Chiriquí | Atlantic | Pacific | E. Foothills | Cana |
|---|---|---|---|---|---|---|---|---|---|---|---|---|---|
| Black-crowned Antpitta | + | + | | | R | | | | R | | | R | U |
| Scaled Antpitta | + | + | | R | R | | R | | R | | | | R |
| Spectacled Antpitta | + | + | | | R | C | | C | | R | R | | R |
| Fulvous-bellied Antpitta | + | + | | | | U | | | | | | | U |
| Ochre-breasted Antpitta | + | + | | | | R | | | R | | | | R |
| Silvery-fronted Tapaculo | + | + | | C | U | | C | | U | | | | |
| Tacarcuna Tapaculo | | + | | | | | | | | | | | |
| Nariño Tapaculo (unnamed sp.) | | + | | | | | | | | | | | R |
| White-fronted Tyrannulet | + | + | | | R | | R | | R | | | | |
| Sooty-headed Tyrannulet | | + | | | | | | | | | | | U |
| Paltry Tyrannulet | + | + | R | C | C | C | U | C | C | C | U | C | U |
| Yellow-bellied Tyrannulet | + | | | | | | | U | | | | | |
| Brown-capped Tyrannulet | + | + | | | | U | | | | C | U | U | C |
| Northern Beardless-Tyrannulet | + | | C | | | | | | | | | | |
| Southern Beardless-Tyrannulet | + | + | R | | | | | U | | U | U | U | |
| Mouse-colored Tyrannulet | | + | | | | | | | | | R | | |
| Northern Scrub-Flycatcher | + | + | R | | | | | R | | R | U | | |
| Yellow-crowned Tyrannulet | + | + | | | | | | R | | C | U | | |
| Forest Elaenia | | + | | | | | | | | U | R | | U |
| Gray Elaenia | | + | | | | | | | | R | | | U |
| Greenish Elaenia | + | + | C | | | | | R | R | U | U | | U |
| Yellow-bellied Elaenia | + | + | U | C | U | C | | C | C | C | C | U | U |
| Lesser Elaenia | + | + | | | | | | U | U | R | U | U | |
| Mountain Elaenia | + | + | | C | C | | C | | C | | | | |
| Torrent Tyrannulet | + | + | | R | C | | | | U | | | | |
| Olive-striped Flycatcher | + | + | | C | C | U | | | U | R | | R | C |
| Ochre-bellied Flycatcher | + | + | | | C | C | | C | R | C | C | C | C |
| Sepia-capped Flycatcher | + | + | | | | R | | | | X | R | | |
| Slaty-capped Flycatcher | + | + | | | U | | | | U | | | | U |
| Yellow Tyrannulet | + | + | | | U | C | | C | R | U | U | U | |
| Yellow-green Tyrannulet | | + | | | | | | | | R | R | | R |
| Rufous-browed Tyrannulet | + | + | | | R | | | | | | | | |
| Bronze-olive Pygmy-Tyrant | | + | | | | | | | | | | | R |
| Black-capped Pygmy-Tyrant | + | + | | | | C | | | | U | | U | U |
| Scale-crested Pygmy-Tyrant | + | + | | C | C | | | C | R | | | C | C |
| Pale-eyed Pygmy-Tyrant | | + | | | | | | | | | R | | |
| Northern Bentbill | + | + | U | | U | C | | C | | | | | |
| Southern Bentbill | | + | | | | | | | | C | C | | C |

|  | Costa Rica | Panama | Guanacaste | Monteverde | Braulio | La Selva | Cerro | Corcovado | Chiriquí | Atlantic | Pacific | E. Foothills | Cana |
|---|---|---|---|---|---|---|---|---|---|---|---|---|---|
| Slate-headed Tody-Flycatcher | + | + | R |  |  | U |  | U |  | R | R |  |  |
| Common Tody-Flycatcher | + | + | C | R | C | C |  | C | U | C | C | U | C |
| Black-headed Tody-Flycatcher | + | + |  |  | U | C |  |  |  | U | R | R | U |
| Brownish Twistwing |  | + |  |  |  |  |  |  |  | U | U |  | U |
| Eye-ringed Flatbill | + | + |  | C | C | U |  | C | R |  |  |  | U |
| Olivaceous Flatbill |  | + |  |  |  |  |  |  |  | U | U | U | C |
| Yellow-breasted Flycatcher |  | + |  |  |  |  |  |  |  |  |  |  |  |
| Yellow-olive Flycatcher | + | + | C | X | U | C |  | C | R |  | U | U |  |
| Yellow-margined Flycatcher | + | + |  |  | U | C |  |  |  | C | U | U | U |
| Stub-tailed Spadebill | + | o | R |  |  |  |  |  |  |  |  |  |  |
| White-throated Spadebill | + | + |  | C | U |  |  |  | R |  |  |  | U |
| Golden-crowned Spadebill | + | + |  |  | U | C |  | C |  | U | R |  | U |
| Royal Flycatcher | + | + | R |  |  | R |  | U | R | U | U | R | U |
| Ruddy-tailed Flycatcher | + | + |  |  | U | C |  | C |  | C | C | C | U |
| Tawny-breasted Flycatcher |  | + |  |  |  |  |  |  |  |  |  |  |  |
| Sulphur-rumped Flycatcher | + | + |  |  | U | U |  | C |  | U | U | U | C |
| Black-tailed Flycatcher | + | + | R |  |  |  |  | U |  | U | U | R | U |
| Bran-colored Flycatcher | + | + |  |  |  |  |  |  | R | R | R |  |  |
| Tawny-chested Flycatcher | + |  |  |  |  | R |  |  |  |  |  |  |  |
| Black-billed Flycatcher |  | + |  |  |  |  |  |  |  |  |  |  | R |
| Tufted Flycatcher | + | + |  |  | U | C |  | C |  | C |  |  | U |
| Olive-sided Flycatcher | + | + | R | U | U | R | U | U | U | R | R | U | U |
| Dark Pewee | + | + |  |  | R | U |  | U |  | U |  |  |  |
| Ochraceous Pewee | + | + |  |  |  |  |  | U |  | R |  |  |  |
| Western Wood-Pewee | + | + | R | U | U | R | U | R | C | R | R | C |  |
| Eastern Wood-Pewee | + | + | U | U | U | C | R | C | R | C | C | U | U |
| Tropical Pewee | + | + | R |  | U | C |  | C |  | U | U | U |  |
| Yellow-bellied Flycatcher | + | + | U | U | R | C |  | C | U | X |  |  | X |
| Acadian Flycatcher | + | + |  | R | U | U |  |  | R | U | U | R | R |
| Willow Flycatcher | + | + | R | U |  | R |  | R | R | R | U | R | R |
| Alder Flycatcher | + | + | U | U | U | U |  | R | R | R | R | R | R |
| White-throated Flycatcher | + | + |  |  |  | R |  |  | R | X |  |  |  |
| Least Flycatcher | + | o | R | R |  | R |  | R |  | X | X |  |  |
| Yellowish Flycatcher | + | + |  | C | U |  | U |  | C |  |  |  |  |
| Black-capped Flycatcher | + | + |  |  |  |  | C |  | U |  |  |  |  |
| Black Phoebe | + | + |  |  | U |  |  |  | C |  |  |  | R |
| Vermilion Flycatcher |  | o |  |  |  |  |  |  | X | X |  |  |  |
| Pied Water-Tyrant |  | + |  |  |  |  |  |  |  | R | U |  |  |

| | Costa Rica | Panama | Guanacaste | Monteverde | Braulio | La Selva | Cerro | Corcovado | Chiriquí | Atlantic | Pacific | E. Foothills | Cana |
|---|---|---|---|---|---|---|---|---|---|---|---|---|---|
| Long-tailed Tyrant | + | + | | | | U | C | | | C | C | C | C |
| Cattle Tyrant | | o | | | | | | | | | X | | X |
| Bright-rumped Attila | + | + | C | C | C | U | | C | U | C | C | U | C |
| Speckled Mourner | + | + | | | | R | | | | R | R | R | R |
| Rufous Mourner | + | + | | R | U | C | | C | | U | | U | C |
| Sirystes | | + | | | | | | | | R | | | U |
| Dusky-capped Flycatcher | + | + | C | C | C | U | | C | U | C | U | C | C |
| Panama Flycatcher | + | + | | | | | | U | R | C | C | U | |
| Nutting's Flycatcher | + | | U | R | | | | | | | | | |
| Great Crested Flycatcher | + | + | C | | C | C | | C | | C | C | U | U |
| Brown-crested Flycatcher | + | | C | | | | | | | | | | |
| Lesser Kiskadee | | + | | | | | | | | U | U | | |
| Great Kiskadee | + | + | C | U | C | C | | U | R | C | C | R | |
| Boat-billed Flycatcher | + | + | C | C | C | C | | C | C | C | C | C | U |
| Rusty-margined Flycatcher | | + | | | | | | | | C | C | | C |
| Social Flycatcher | + | + | C | C | C | C | | C | C | C | C | C | |
| Gray-capped Flycatcher | + | + | | | U | C | | C | U | U | R | | C |
| White-ringed Flycatcher | + | + | | | | U | | | | U | R | R | U |
| Golden-bellied Flycatcher | + | + | | C | C | | U | | U | | | | |
| Golden-crowned Flycatcher | | + | | | | | | | | | | | U |
| Streaked Flycatcher | + | + | U | | U | X | | U | C | C | C | C | C |
| Sulphur-bellied Flycatcher | + | + | C | C | | U | | U | | R | R | U | |
| Piratic Flycatcher | + | + | C | | U | U | | C | C | C | C | C | C |
| Tropical Kingbird | + | + | C | C | C | C | U | C | C | C | C | C | C |
| Western Kingbird | + | o | U | | | | | | | | X | | |
| Eastern Kingbird | + | + | U | R | U | C | | U | | C | C | C | U |
| Gray Kingbird | o | + | | | | | | | | U | R | | |
| Scissor-tailed Flycatcher | + | + | C | | | | | U | | | X | | |
| Fork-tailed Flycatcher | + | + | R | | | | | R | U | U | C | U | |
| Barred Becard | + | + | | R | R | | | | R | | | | |
| Cinereous Becard | | + | | | | | | | | X | X | | R |
| White-winged Becard | + | + | R | | U | C | | C | U | U | U | R | R |
| Cinnamon Becard | + | + | | R | R | C | | U | | R | R | U | C |
| Black-and-White Becard | + | + | | R | R | | | | R | | | | |
| Rose-throated Becard | + | + | U | | | X | | | R | | | | |
| One-colored Becard | | + | | | | | | | | X | | | C |
| Masked Tityra | + | + | C | C | C | C | | C | U | C | C | U | C |
| Black-crowned Tityra | + | + | U | | U | U | | U | R | U | U | U | U |

| | Costa Rica | Panama | Guanacaste | Monteverde | Braulio | La Selva | Cerro | Corcovado | Chiriqui | Atlantic | Pacific | E. Foothills | Cana |
|---|---|---|---|---|---|---|---|---|---|---|---|---|---|
| Rufous Piha | + | + | | | U | C | | C | | U | R | U | U |
| Turquoise Cotinga | + | + | | | | | | U | R | | | | |
| Lovely Cotinga | + | o | | | | | | | | | | | |
| Blue Cotinga | | + | | | | | | | | U | R | U | U |
| Black-tipped Cotinga | | + | | | | | | | | | | | U |
| Yellow-billed Cotinga | + | o | | | | | | U | | | | | |
| Snowy Cotinga | + | + | | | | U | | | | | | | |
| Purple-throated Fruitcrow | + | + | | | | C | | | | C | C | C | C |
| Bare-necked Umbrellabird | + | + | | R | R | R | | | R | | | | |
| Three-wattled Bellbird | + | + | | C | U | R | C | | U | X | | | |
| Thrushlike Mourner | + | + | | | | R | | U | | R | R | U | C |
| Broad-billed Sapoya | | + | | | | | | | | R | | X | R |
| Gray-headed Piprites | + | | | | | R | | | | | | | |
| Green Manakin | | + | | | | | | | | | R | | |
| White-collared Manakin | + | + | | | U | C | | | | | | | |
| Golden-collared Manakin | | + | | | | | | | | C | C | C | C |
| Orange-collared Manakin | + | + | | | | | | C | | | | | |
| White-ruffed Manakin | + | + | | | C | U | | R | R | | | C | C |
| Lance-tailed Manakin | + | + | | | | | | | | R | R | U | |
| Long-tailed Manakin | + | | C | C | | | | | | | | | |
| White-crowned Manakin | + | + | | | | U | | | R | | | | |
| Blue-crowned Manakin | + | + | | | | | | C | | C | U | C | C |
| Golden-headed Manakin | | + | | | | | | | | | | | C |
| Red-capped Manakin | + | + | | | C | C | | C | | C | C | U | |
| Sharpbill | + | + | | | R | | | | X | | | | U |
| Purple Martin | + | + | | R | | | | R | | R | R | | |
| Gray-breasted Martin | + | + | C | R | | U | | C | U | C | C | C | U |
| Southern Martin | | o | | | | | | | X | | | | |
| Brown-chested Martin | o | + | | | | | | | | R | R | | R |
| Tree Swallow | + | + | R | | | | | | | R | R | | |
| Mangrove Swallow | + | + | C | | | C | | C | | C | C | | |
| Violet-green Swallow | + | o | R | X | | | X | X | | | | | |
| Blue-and-White Swallow | + | + | | C | C | | C | | C | R | | | X |
| White-thighed Swallow | | + | | | | | | | | R | | | R |
| N. Rough-winged Swallow | + | + | C | C | C | U | | U | R | C | C | | |
| S. Rough-winged Swallow | + | + | C | U | C | C | | C | C | C | C | C | C |

| | Costa Rica | Panama | Guanacaste | Monteverde | Braulio | La Selva | Cerro | Corcovado | Chiriquí | Atlantic | Pacific | E. Foothills | Cana |
|---|---|---|---|---|---|---|---|---|---|---|---|---|---|
| Bank Swallow | + | + | U | R | | R | | U | | U | U | | |
| Cliff Swallow | + | + | U | R | | U | | C | R | U | U | R | R |
| Cave Swallow | | o | | | | | | | | | X | | |
| Barn Swallow | + | + | C | R | U | C | R | C | C | C | C | C | R |
| White-throated Magpie-Jay | + | | C | R | | | | | | | | | |
| Black-chested Jay | + | + | | | | | | | U | C | U | U | C |
| Brown Jay | + | + | R | C | U | U | | | X | | | | |
| Azure-hooded Jay | + | + | | C | U | | U | | R | | | | |
| Silvery-throated Jay | + | + | | | | | R | | R | | | | |
| Black-capped Donocobius | | + | | | | | | | | | | | R |
| White-headed Wren | | + | | | | | | | | R | | | C |
| Band-backed Wren | + | + | | | C | C | | | | | | | |
| Rufous-naped Wren | + | | C | | | | | | | | | | |
| Rock Wren | + | | R | | | | | | | | | | |
| Sooty-headed Wren | | + | | | | | | | | | | | U |
| Black-throated Wren | + | + | | | U | U | | | | | | | |
| Black-bellied Wren | + | + | | | | | | C | | C | C | C | C |
| Bay Wren | + | + | | | C | C | | | | C | R | U | C |
| Riverside Wren | + | + | | | | | | C | R | | | | |
| Stripe-throated Wren | | + | | | | | | | | | | R | U |
| Stripe-breasted Wren | + | + | | | U | C | | | R | | | | |
| Rufous-and-White Wren | + | + | C | C | | | | | R | U | U | | |
| Rufous-breasted Wren | + | + | | R | | | | U | U | U | U | | |
| Banded Wren | + | | C | | | | | | | | | | |
| Buff-breasted Wren | | + | | | | | | | | C | C | | |
| Plain Wren | + | + | U | C | | R | | C | C | C | C | U | |
| House Wren | + | + | U | C | U | U | U | C | C | C | C | U | C |
| Ochraceous Wren | + | + | | C | C | | C | | U | | | | C |
| Sedge Wren | + | o | | | | | | | X | | | | |
| Timberline Wren | + | + | | | | | C | | U | | | | |
| White-breasted Wood-Wren | + | + | | | U | C | C | | U | R | C | C | C |
| Gray-breasted Wood-Wren | + | + | | | C | C | | C | C | | | C | C |
| Northern Nightingale-Wren | + | | | | | U | U | | | | | | |
| Southern Nightingale-Wren | + | + | | | | | | C | R | U | U | R | R |
| Song Wren | + | + | | | U | U | | | | C | C | U | C |
| N. American Dipper | + | + | | | U | | U | | U | | | | |
| Tawny-faced Gnatwren | + | + | | | U | U | | | | U | U | C | U |

| | Costa Rica | Panama | Guanacaste | Monteverde | Braulio | La Selva | Cerro | Corcovado | Chiriqui | Atlantic | Pacific | E. Foothills | Cana |
|---|---|---|---|---|---|---|---|---|---|---|---|---|---|
| Long-billed Gnatwren | + | + | C | | U | C | | C | R | C | C | U | R |
| White-lored Gnatcatcher | + | | C | | | | | | | | | | |
| Tropical Gnatcatcher | + | + | C | | C | C | | C | R | C | C | U | |
| Slate-throated Gnatcatcher | | + | | | | | | | | | | | U |
| Black-faced Solitaire | + | + | | C | C | | C | | U | | | | |
| Varied Solitaire | | + | | | | | | | | | | | C |
| Black-bld Nightingale-Thrush | + | + | | | | | C | | C | | | | |
| Orange-billed N.-Thrush | + | + | | C | | | | | U | | | | |
| Ruddy-capped N.-Thrush | + | + | | C | C | | C | | C | | | | |
| Slaty-backed N.-Thrush | + | + | | C | U | | | | U | | | R | U |
| Black-headed N.-Thrush | + | + | | C | C | | | | | | | | |
| Veery | + | + | | | | | R | | | X | R | X | R |
| Gray-cheeked Thrush | + | + | R | R | U | U | | U | R | R | R | R | R |
| Swainson's Thrush | + | + | C | C | C | C | R | C | C | C | C | C | C |
| Wood Thrush | + | + | U | R | C | C | | U | R | R | | R | |
| Sooty Thrush | + | + | | | | | C | | U | | | | |
| Mountain Thrush | + | + | | C | C | | C | | C | | | | |
| Pale-vented Thrush | + | + | | | U | | | | | | | R | U |
| Clay-colored Thrush | + | + | C | C | C | C | R | C | C | C | C | C | |
| White-throated Thrush | + | + | | C | | | | U | R | R | R | U | U |
| Gray Catbird | + | + | | | U | U | | | R | U | U | R | |
| Tropical Mockingbird | | + | | | | | | | | C | C | | |
| Yellowish Pipit | | + | | | | | | | | U | R | | |
| Cedar Waxwing | + | + | X | R | R | | | | R | R | R | | |
| Blk.-and-Yel. Silky-Flycatcher | + | + | | C | R | | C | | U | | | | |
| Long-tailed Silky-Flycatcher | + | + | | | | | C | | C | | | | |
| White-eyed Vireo | o | o | | | | | | | | | X | | |
| Mangrove Vireo | + | | | | | | | | | | | | |
| Solitary Vireo | o | o | X | X | | | X | | X | X | | | |
| Yellow-throated Vireo | + | + | C | R | U | C | | C | U | U | U | C | U |
| Yellow-winged Vireo | + | + | | | | | C | | C | | | | |
| Warbling Vireo | o | o | | | | | | | | | | | |
| Brown-capped Vireo | + | + | | | C | U | U | | C | | | | |
| Philadelphia Vireo | + | + | U | U | U | U | | C | U | R | R | | |
| Red-eyed Vireo | + | + | U | C | C | C | | U | C | C | C | C | C |
| Yellow-green Vireo | + | + | C | U | U | U | | U | U | C | C | U | U |
| Black-whiskered Vireo | o | o | | | | | | | | R | | | |

| | Costa Rica | Panama | Guanacaste | Monteverde | Braulio | La Selva | Cerro | Corcovado | Chiriquí | Atlantic | Pacific | E. Foothills | Cana |
|---|---|---|---|---|---|---|---|---|---|---|---|---|---|
| Scrub Greenlet | + | + | | | | | | U | R | C | C | | |
| Tawny-crowned Greenlet | + | + | | U | U | R | | C | R | R | R | R | U |
| Golden-fronted Greenlet | | + | | | | | | | | U | U | | |
| Lesser Greenlet | + | + | C | C | C | C | | C | U | C | C | C | C |
| Green Shrike-Vireo | + | + | | | U | C | | C | | U | U | U | |
| Yellow-browed Shrike-Vireo | | + | | | | | | | | | | | R |
| Rufous-browed Peppershrike | + | + | | C | | | | R | R | | X | | |
| Blue-winged Warbler | + | + | | | | R | | | | U | R | | |
| Golden-winged Warbler | + | + | R | C | U | U | C | U | U | U | U | U | U |
| Tennessee Warbler | + | + | C | C | C | C | R | C | C | C | C | C | C |
| Nashville Warbler | o | o | X | | | | | X | | | | | |
| Northern Parula | o | o | | | | | | | | R | | | |
| Tropical Parula | + | + | | U | C | | U | | C | | | | C |
| Flame-throated Warbler | + | + | | | | | C | | U | | | | |
| Yellow Warbler | + | + | C | R | U | C | | C | R | C | C | C | R |
| Chestnut-sided Warbler | + | + | C | X | C | C | | C | U | C | C | C | U |
| Magnolia Warbler | + | + | R | | | R | | R | | U | U | U | |
| Cape May Warbler | + | + | | X | | | | R | R | R | | | |
| Black-throated Blue Warbler | o | o | | X | | | | X | R | | | | |
| Yellow-rumped Warbler | + | + | X | R | | R | | R | R | U | U | | |
| Townsend's Warbler | + | o | | R | | | R | | X | | | | |
| Hermit Warbler | o | o | | X | X | | | | X | | | | |
| Black-throated Green Warbler | + | + | R | C | C | R | C | | C | R | R | | R |
| Blackburnian Warbler | + | + | R | U | C | U | U | U | C | U | U | | C |
| Yellow-throated Warbler | + | + | | | | | | | | R | R | R | R |
| Palm Warbler | + | + | | | | | | | | X | R | R | |
| Bay-breasted Warbler | + | + | R | | R | U | | U | C | C | C | C | C |
| Blackpoll Warbler | o | o | X | | | | | | | R | R | | |
| Cerulean Warbler | + | + | | R | R | U | | | | R | R | U | R |
| Black-and-White Warbler | + | + | U | U | C | U | | U | C | C | C | C | C |
| American Redstart | + | + | R | | R | U | R | U | U | U | U | U | U |
| Prothonotary Warbler | + | + | C | | | U | | U | R | C | C | R | R |
| Worm-eating Warbler | + | + | R | U | R | U | | R | R | R | R | R | |
| Ovenbird | + | + | C | U | U | C | | C | R | R | | R | |
| Northern Waterthrush | + | + | C | R | | C | | C | U | C | C | R | |
| Louisiana Waterthrush | + | + | R | R | U | U | U | R | R | R | R | R | R |
| Kentucky Warbler | + | + | R | C | U | C | | U | R | C | C | R | R |

|  | Costa Rica | Panama | Guanacaste | Monteverde | Braulio | La Selva | Cerro | Corcovado | Chiriquí | Atlantic | Pacific | E. Foothills | Cana |
|---|---|---|---|---|---|---|---|---|---|---|---|---|---|
| Connecticut Warbler | o | o |  |  |  |  | X |  |  |  |  |  |  |
| Mourning Warbler | + | + | U | U | U | C |  | C | C | C | C | U | C |
| MacGillivray's Warbler | + | + | R | R |  | R |  |  | R |  |  |  |  |
| Common Yellowthroat | + | + | R |  |  | X |  |  | R | U | U |  |  |
| Olive-crowned Yellowthroat | + | + |  | R | U | C |  |  |  |  |  |  |  |
| Masked Yellowthroat | + | + |  |  |  |  |  |  | U |  |  |  |  |
| Gray-crowned Yellowthroat | + | + | C | U |  |  |  | C | U |  |  |  |  |
| Hooded Warbler | + | + |  |  |  | R |  |  |  | R | R |  |  |
| Wilson's Warbler | + | + | R | C | C | C | C | R | C |  | R |  |  |
| Canada Warbler | + | + | U | U | U | C |  | U | U | C | C | C | U |
| Slate-throated Whitestart | + | + |  | C | C |  | C |  | C |  |  | R | U |
| Collared Whitestart | + | + |  | C | U |  | C |  | C |  |  |  |  |
| Golden-crowned Warbler | + | + |  | C | U |  |  |  | U |  |  |  |  |
| Rufous-capped Warbler | + | + | C | U |  |  |  |  | U | U | U | C |  |
| Black-cheeked Warbler | + | + |  |  |  |  | C |  | C |  |  |  |  |
| Pirré Warbler |  | + |  |  |  |  |  |  |  |  |  |  | U |
| Three-striped Warbler | + | + |  | C | C |  | C |  | R |  |  | R |  |
| Buff-rumped Warbler | + | + |  | X | U | C |  | C | R | R | R |  | U |
| Wrenthrush | + | + |  | U |  |  | U |  | R |  |  |  |  |
| Yellow-breasted Chat | + | + |  |  |  | U |  | R |  |  |  |  |  |
| Bananaquit | + | + |  | C | C | C |  | C |  | C | C | C | U |
| White-eared Conebill |  | + |  |  |  |  |  |  |  |  |  |  | R |
| Plain-colored Tanager | + | + |  |  |  | C |  |  |  | C | C | U | U |
| Gray-and-Gold Tanager |  | + |  |  |  |  |  |  |  |  |  | X | R |
| Emerald Tanager | + | + |  |  | C |  |  |  | R |  |  | C | U |
| Silver-throated Tanager | + | + |  | C | C | U | C | U | C |  |  | U | U |
| Speckled Tanager | + | + |  |  | C |  |  |  | R |  |  | U | U |
| Bay-headed Tanager | + | + |  |  | C | X |  | C | U | U | U | C | C |
| Rufous-winged Tanager | + | + |  |  | R |  |  |  |  |  |  | U |  |
| Golden-hooded Tanager | + | + |  |  | U | C |  | C | U | C | C | C | C |
| Spangle-cheeked Tanager | + | + |  | C | C |  | C |  | U |  |  |  |  |
| Green-naped Tanager |  | + |  |  |  |  |  |  |  |  |  |  | U |
| Scarlet-thighed Dacnis | + | + |  | C | C | R |  | U |  | U | U | C | U |
| Blue Dacnis | + | + |  |  | C | U |  | C |  | C | C | C | C |
| Viridian Dacnis |  | + |  |  |  |  |  |  |  |  |  |  | R |
| Green Honeycreeper | + | + |  |  | U | U |  | R |  | C | C | C | C |
| Shining Honeycreeper | + | + |  |  | C | C |  | C |  |  | U | C | U |

| | Costa Rica | Panama | Guanacaste | Monteverde | Braulio | La Selva | Cerro | Corcovado | Chiriquí | Atlantic | Pacific | E. Foothills | Cana |
|---|---|---|---|---|---|---|---|---|---|---|---|---|---|
| Purple Honeycreeper | | + | | | | | | | | | | | U |
| Red-legged Honeycreeper | + | + | C | U | | | | C | | C | C | C | |
| Golden-browed Chlorophonia | + | + | | C | C | | U | | U | | | | |
| Yellow-collared Chlorophonia | | + | | | | | | | | | | | X |
| Scrub Euphonia | + | | C | X | | | | | | | | | |
| Yellow-crowned Euphonia | + | + | | R | | C | | C | | U | C | | |
| Thick-billed Euphonia | + | + | | | | | | C | R | C | C | C | |
| Yellow-throated Euphonia | + | o | U | U | | | | | | | | | |
| Blue-hooded Euphonia | + | + | | C | R | | U | | U | | | | |
| Fulvous-vented Euphonia | | + | | | | | | | | U | U | U | U |
| Spot-crowned Euphonia | + | + | | | | | | | R | | | | |
| Olive-backed Euphonia | + | + | | | C | C | | | | | | | |
| White-vented Euphonia | + | + | | | U | U | | U | | R | R | C | |
| Tawny-capped Euphonia | + | + | | | U | | | | U | | U | | |
| Orange-bellied Euphonia | | + | | | | | | | | | | | U |
| Blue-gray Tanager | + | + | C | C | C | C | | C | C | C | C | C | C |
| Palm Tanager | + | + | | X | C | C | | C | C | C | C | C | C |
| Blue-and-Gold Tanager | + | + | | | U | | | | U | | | | |
| Olive Tanager | + | + | | | C | C | | | | R | R | U | |
| Lemon-spectacled Tanager | | + | | | | | | | | | | | U |
| Gray-headed Tanager | + | + | U | | | | | C | R | U | U | R | R |
| White-throated Shrike-Tanager | + | + | | | | R | | U | R | | | | |
| Sulphur-rumped Tanager | + | + | | | | | | | | U | | | |
| Scarlet-browed Tanager | | + | | | | | | | | | | | U |
| White-shouldered Tanager | + | + | | | | U | | C | | C | C | C | C |
| Tawny-crested Tanager | + | + | | | C | C | | | | | R | C | |
| White-lined Tanager | + | + | | | C | C | | U | | | | | |
| Red-crowned Ant-Tanager | + | + | R | | | | | | R | | R | U | |
| Red-throated Ant-Tanager | + | + | | | C | C | | | | C | C | | |
| Black-cheeked Ant-Tanager | + | | | | | | | U | | | | | |
| Hepatic Tanager | + | + | | C | C | | | U | U | | U | U | |
| Summer Tanager | + | + | C | C | C | C | | C | C | C | C | C | C |
| Scarlet Tanager | + | + | U | U | U | U | | U | U | U | U | U | U |
| Western Tanager | + | + | C | | | | | R | R | | | | |
| Flame-colored Tanager | + | + | | | | U | | U | | | | | |
| White-winged Tanager | + | + | | R | U | | | | U | | | | |

|  | Costa Rica | Panama | Guanacaste | Monteverde | Braulio | La Selva | Cerro | Corcovado | Chiriquí | Atlantic | Pacific | E. Foothills | Cana |
|---|---|---|---|---|---|---|---|---|---|---|---|---|---|
| Crimson-collared Tanager | + | + |  |  | X | U | U |  |  |  |  |  |  |
| Crimson-backed Tanager |  | + |  |  |  |  |  |  |  | C | C |  | C |
| Scarlet-rumped Tanager | + | + |  |  |  | C | C |  | C | U |  |  |  |
| Flame-rumped Tanager |  | + |  |  |  |  |  |  |  | C | R |  | C |
| Rosy Thrush-Tanager | + | + |  |  |  |  |  |  |  | U | U |  |  |
| Dusky-faced Tanager | + | + |  |  |  | U | C |  |  | U |  | U | U |
| Common Bush-Tanager | + | + |  | X | C | C |  | C |  | C |  |  |  |
| Tacarcuna Bush-Tanager |  | + |  |  |  |  |  |  |  |  |  | U |  |
| Pirré Bush-Tanager |  | + |  |  |  |  |  |  |  |  |  |  | U |
| Sooty-capped Bush-Tanager | + | + |  |  | C |  |  | C | C |  |  |  |  |
| Yellow-throated Bush-Tanager |  | + |  |  |  |  |  |  |  |  |  |  |  |
| Ashy-throated Bush-Tanager | + | o |  |  |  | U |  |  | X |  |  |  |  |
| Yellow-backed Tanager |  | + |  |  |  |  |  |  |  |  |  |  | U |
| Black-and-Yellow Tanager | + | + |  |  |  | U |  |  | R |  |  | U | U |
| Swallow Tanager |  | + |  |  |  |  |  |  |  |  |  |  | C |
| Streaked Saltator | + | + |  |  |  |  |  |  | C | C | C | U |  |
| Grayish Saltator | + |  | R | R |  |  |  |  |  |  |  |  |  |
| Buff-throated Saltator | + | + | R | C | C | C |  | C | C | C | C | U | C |
| Black-headed Saltator | + | + |  |  | U | C |  |  | X | U |  | R | X |
| Slate-colored Grosbeak | + | + |  |  | R | U | U |  |  | U | R | U | U |
| Black-faced Grosbeak | + | + |  |  | R | U | C |  |  | X |  |  |  |
| Yellow-green Grosbeak |  | + |  |  |  |  |  |  |  |  |  |  | C |
| Black-thighed Grosbeak | + | + |  |  | U | U |  | U |  |  |  |  |  |
| Rose-breasted Grosbeak | + | + | U | R | U | C |  | U | C | R | R |  | U |
| Black-headed Grosbeak | o |  |  |  |  |  |  |  |  |  |  |  |  |
| Blue-black Grosbeak | + | + |  |  | C | C |  | C | R | C | U | U | C |
| Blue Grosbeak | + | + | U |  |  |  |  | R |  | R | R |  |  |
| Indigo Bunting | + | + | U | R |  | R |  | U | U | R | R |  |  |
| Painted Bunting | + | + | U |  |  |  |  | R | R |  |  |  |  |
| Dickcissel | + | + | U |  |  |  |  | U | R | U | U |  |  |
| Sooty-faced Finch | + | + |  |  | U | U |  | R |  |  |  |  |  |
| Yellow-thighed Finch | + | + |  |  | C |  | C | C |  |  |  |  |  |
| Yellow-green Finch |  | + |  |  |  |  |  | U |  |  |  |  |  |
| Large-footed Finch | + | + |  |  |  |  | C | U |  |  |  |  |  |
| Yellow-throated Brush-Finch | + | + |  |  | C |  |  | C |  |  |  |  |  |
| Chestnut-capped Brush-Finch | + | + |  |  | C | U |  | U |  |  |  |  | U |
| Black-headed Brush-Finch | + | + |  |  |  |  |  | R |  |  |  | R | U |

| | Costa Rica | Panama | Guanacaste | Monteverde | Braulio | La Selva | Cerro | Corcovado | Chiriquí | Atlantic | Pacific | E. Foothills | Cana |
|---|---|---|---|---|---|---|---|---|---|---|---|---|---|
| Orange-billed Sparrow | + | + | | | C | C | | C | R | C | U | U | C |
| Olive Sparrow | + | | C | | | | | | | | | | |
| Black-striped Sparrow | + | + | | | U | C | | C | U | C | C | U | C |
| Prevost's Ground-Sparrow | + | | | | | | | | | | | | |
| White-eared Ground-Sparrow | + | | | C | | | | | | | | | |
| Blue-black Grassquit | + | + | C | | C | C | | C | C | C | C | C | |
| Slate-colored Seedeater | + | + | | | R | | | R | R | R | R | R | R |
| Variable Seedeater | + | + | | R | C | C | | C | C | C | C | C | C |
| White-collared Seedeater | + | + | C | | | U | | U | U | | | | |
| Lesson's Seedeater | | o | | | | | | | | | | | R |
| Yellow-bellied Seedeater | + | + | | | | | | U | U | U | U | U | U |
| Ruddy-breasted Seedeater | + | + | | | | | | R | R | | U | | |
| Nicaraguan Seed-Finch | + | + | | | | U | | | | | | | |
| Lesser Seed-Finch | + | + | | | | C | | C | U | C | C | U | U |
| Blue Seedeater | + | + | | R | | | R | | R | | R | R | |
| Yellow-faced Grassquit | + | + | | C | C | C | | U | C | R | R | C | |
| Slaty Finch | + | + | | R | | | | | R | | | | |
| Slaty Flower-piercer | + | + | | C | | | C | | C | | | | |
| Peg-billed Finch | + | o | | R | | | R | | X | | | | |
| Saffron Finch | | + | | | | | | | | | C | | |
| Grassland Yellow-Finch | o | + | X | | | | | | | | | | |
| Wedge-tailed Grass-Finch | + | + | | | | | | | X | | | | |
| Stripe-headed Sparrow | + | | C | X | | | | | | | | | |
| Botteri's Sparrow | + | | R | | | | | | | | | | |
| Rusty Sparrow | + | | R | | | | | | | | | | |
| Chipping Sparrow | o | | | | | | | | | | | | |
| Grasshopper Sparrow | + | + | R | | | | | | | | | | |
| Lark Sparrow | | o | | | | | | | | | X | | |
| Savannah Sparrow | o | | | | | | | | | | | | |
| Lincoln's Sparrow | + | o | | R | | | | | | | | | |
| Rufous-collared Sparrow | + | + | | C | C | | C | | C | | | | |
| White-crowned Sparrow | | o | | | | | | | | | X | | |
| Volcano Junco | + | + | | | | | U | | U | | | | |
| Bobolink | + | + | | | | | | | | R | R | | |
| Red-winged Blackbird | + | | U | | | | | | | | | | |
| Red-breasted Blackbird | + | + | | | | | | | R | U | U | | R |
| Eastern Meadowlark | + | + | C | C | | C | U | | C | U | U | U | |

| | Costa Rica | Panama | Guanacaste | Monteverde | Braulio | La Selva | Cerro | Corcovado | Chiriquí | Atlantic | Pacific | E. Foothills | Cana |
|---|---|---|---|---|---|---|---|---|---|---|---|---|---|
| Yellow-headed Blackbird | o | o | X | | | | | | | X | X | | |
| Melodious Blackbird | o | | X | | | | | | | | | | |
| Nicaraguan Grackle | + | | | | | | | | | | | | |
| Great-tailed Grackle | + | + | C | C | | R | | C | | C | C | | |
| Shiny Cowbird | | + | | | | | | | | R | R | | R |
| Bronzed Cowbird | + | + | U | U | | R | | C | U | R | U | | |
| Giant Cowbird | + | + | | | U | U | | U | | U | U | | U |
| Black-cowled Oriole | + | + | | | U | U | | | | | | | |
| Orchard Oriole | + | + | U | | | U | | U | U | C | C | U | |
| Yellow-backed Oriole | | + | | | | | | | | C | C | U | U |
| Orange-crowned Oriole | | + | | | | | | | | R | | R | |
| Yellow-tailed Oriole | + | + | | | R | | | | | U | R | | U |
| Streak-backed Oriole | + | | C | | | | | | | | | | |
| Spot-breasted Oriole | + | | U | | | | | | | | | | |
| Northern Oriole | + | + | C | C | C | C | | C | C | C | C | C | C |
| Yellow-billed Cacique | + | + | | | U | C | R | C | | U | U | | C |
| Scarlet-rumped Cacique | + | + | | | C | C | | C | | C | C | C | |
| Yellow-rumped Cacique | | + | | | | | | | | C | C | | C |
| Crested Oropendola | | + | | | | | | | | U | U | U | C |
| Chestnut-headed Oropendola | + | + | | R | C | U | | R | U | C | C | U | C |
| Montezuma Oropendola | + | + | | R | C | C | | | | U | | | |
| Black Oropendola | | + | | | | | | | | | | | |
| Yellow-bellied Siskin | + | + | | | U | | U | | C | | | | |
| Lesser Goldfinch | + | + | | | | | | | C | | U | U | |
| House Sparrow | + | + | U | | | | | | R | R | U | | |

# Index of Locations

**A**
Achiote Road 108
Aguadulce 122
**B**
Bagaces 46
Barra Honda National Park 50
Bay of Panama 99
Bocas del Toro 137
Boquete 127
Braulio Carrillo National Park 54
**C**
Cahuita 82
Cana 116
Canal Zone 99
Caribbean lowlands 78
Cerro Azul 114
Cerro Campana 118
Cerro Colorado 134
Cerro de la Muerte 67
Cerro Jefe 114
Cerro Punta 128
Chiriquí 126
Chiriquí Grande 137
Chirripó National Park 54
Chiva Chiva Road 100
Chorcha Abajo 134
Colón 107
**D**
David 126
**E**
El Copé 119
El Real 116
Escobal Road 108
**F**
Finca La Selva 78
Finca Lérida 127
Fort San Lorenzo 109
Fort Sherman 109
Fortuna 131
Free Zone 108
**G**
Galeta Road 109
Gamboa 100
Gatun Dam 108

Genesis II 68
Golfito 37
Golfito National Wildlife Refuge 37
Guanacaste 42
Guanacaste Conservation Area 44
Guanacaste National Park 42
Guayabo National Monument 61
**I**
INRENARE office 103
**J**
Juan Diaz 99
**L**
La Amistad National Park 54
La Selva 78
La Trinidad 68
Las Cruces Botanical Gardens 70
Las Lajas 136
Liberia 44
Limbo Camp 104
Limón 82
Lomas de Barbudal Biological Reserve 42
**M**
Madden Dam 101
Madden Forest 100
Manzanillo 84
Metropolitan Nature Park 99
Monteverde Biological Reserve 73
**N**
Nicoya 50
Nicoya Peninsula 51
**O**
Oleoducto Road 137
Orosi 57
**P**
Palo Verde National Park 46
Panama City 99
Panama Viejo 99
Penonomé 119
Pipeline Road 103
Plantation Road 100
Playa Hermosa 42
Playa Naranjo 43
Playas del Coco 42
Prussia 59
Puerto Viejo de Sarapiquí 78
Puerto Viejo de Talamanca 82

**R**
Riande Airport Hotel 113
Rincón de la Vieja National Park 42, 54
Rio Tempisque 46

**S**
S-9 Road 109
Samara 42
Santa Clara 129
Santa Fé 123
Santa Rosa National Park 43
Santiago 123
Summit Gardens 100

**T**
Tamarindo 51
Tapantí National Park 57
The Alamo 104
Tiger Trail 109
Tocumen Marsh 113
Tortuguero National Park 86
Turrialba 61

**V**
Varablanca 65
Virgen del Socorro 65
Volcán 127
Volcán Arenal 54
Volcán Barú 126, 127
Volcán Irazú National Park 59
Volcán Lakes 128
Volcán Poás National Park 63

# Index to Species

**A**
Anhinga 48, 87, 113
Ani
   Greater 109, 114
   Groove-billed 45, 49, 62
   Smooth-billed 32, 34, 39, 114
Antbird
   Bare-crowned 35
   Bicolored 35, 56, 106, 124
   Chestnut-backed 28, 32, 35, 39, 56, 80, 85, 102, 106, 121,
   Dull-mantled 106, 124
   Dusky 35, 39, 80, 85, 102, 106
   Immaculate 66, 75, 121, 124, 132
   Ocellated 106
   Spotted 102, 106
   White-bellied 102
   Wing-banded 106
Antpitta
   Black-crowned 55, 106, 118, 119, 121, 124
   Fulvous-bellied 80
   Spectacled 28, 35, 39, 56, 80, 106, 121, 124
Antshrike
   Barred 28, 45, 49, 62, 87, 102, 114
   Black-hooded 28, 32, 35, 39
   Fasciated 84, 102, 106, 124
   Great 28, 35, 39, 80, 106
   Russet 56, 72, 120, 124, 132
   Slaty 80, 84, 102, 106, 119
Ant-Tanager
   Black-cheeked 36, 37, 40
   Red-crowned 29, 72, 122, 126, 133
   Red-throated 81, 85, 122, 126
Antthrush
   Black-faced 28, 35, 39, 62, 80, 102, 106, 119, 121
   Black-headed 115, 119
   Rufous-breasted 133
Antvireo
   Plain 35, 39, 72, 119, 121, 124
   Spot-crowned 106, 121
   Streak-crowned 56, 80
Antwren
   Checker-throated 80, 85, 102, 106
   Dot-winged 28, 35, 85, 102, 124
   Pygmy 106
   Rufous-rumped 132
   Slaty 35, 58, 62, 66, 72, 75, 121, 124
   Streaked 106
   White-flanked 80, 102, 106, 124
Aracari
   Collared 45, 49, 55, 62, 66, 80, 84, 101, 105, 119
   Fiery-billed 28, 35, 39, 129
Attila
   Bright-rumped 29, 35, 40, 45, 49, 58, 72, 75, 85, 102, 121, 125

**B**
Bananaquit 29, 33, 36, 40, 56, 62, 66, 72, 81, 85, 87, 121, 125, 133
Barbet
   Prong-billed 58, 66, 69, 75, 132, 135
   Red-headed 58, 62, 66, 120, 132
   Spot-crowned 108
Barbtail
   Spotted 56, 58, 66, 75, 120, 132, 135
Barbthroat
   Band-tailed 28, 35, 39, 120, 124, 132
Becard
   Black-and-White 133
   Cinnamon 36, 40, 62, 80, 121, 125
   Rose-throated 46, 49, 130, 134
   White-winged 29, 32, 36, 40, 80, 133
Bellbird
   Three-wattled 29, 75, 121, 130, 133
Bentbill
   Northern 28, 35, 40, 45, 80
   Southern 106
Bittern
   Least 113
Blackbird
   Red-breasted 37, 108, 114
   Red-winged 87
Black-Hawk
   Common 84, 109
   Great 44, 48, 105, 131
   Mangrove 30, 34, 38, 44, 48, 99, 101, 123

173

Bobwhite
    Spot-bellied 44, 48
Booby
    Brown 32, 34, 44
Brilliant
    Green-crowned 58, 62, 66, 74, 132
Brush-Finch
    Black-headed 72, 115
    Chestnut-capped 76, 119, 122,135
    Yellow-throated 59, 63, 76, 131, 134
Bunting
    Indigo 46, 50
Bush-Tanager
    Ashy-throated 55, 56
    Common 59, 66, 69, 72, 76, 122, 126, 131, 134, 135
    Sooty-capped 60, 65, 69, 76, 131, 134, 135
    Tacarcuna 115
    Yellow-throated 122, 126

**C**

Cacique
    Scarlet-rumped 29, 36, 40, 56, 81, 107, 122, 126
    Yellow-billed 29, 36, 40, 81, 87, 107, 126
Caracara
    Crested 28, 44, 48, 101, 113
    Red-throated 34
    Yellow-headed 28, 32, 101, 123
Chachalaca
    Gray-headed 32, 61, 84, 120, 124
    Plain 44, 48, 51
Chlorophonia
    Golden-browed 59, 76, 130, 133
Coquette
    White-crested 35
Cormorant
    Neotropic 30, 48, 87
Cotinga
    Blue 106
    Snowy 80, 85
    Turquoise 36, 40
    Yellow-billed 29, 30, 36, 40
Cowbird
    Giant 63, 85

Crake
    White-throated 28, 34, 38, 48, 71, 79, 84, 87
Cuckoo
    Little 114
    Mangrove 45, 49, 51
    Squirrel 28, 34, 39, 45, 49, 55, 61, 65, 74, 79, 84, 101, 129
    Striped 34, 71, 123
Curassow
    Great 33, 34, 44, 105, 132

**D**

Dacnis
    Blue 29, 36, 40, 72, 115
    Scarlet-thighed 29, 36, 40, 56, 59, 62, 66, 72, 75, 121, 129
Dipper
    N. Am. 58, 66, 130, 133
Dove
    Gray-chested 28, 38, 55, 79, 84, 105
    Inca 28, 45, 48, 51
    Mourning 60, 119
    White-tipped 28, 32, 34, 38, 45, 48, 61, 74, 84, 87, 101, 135
    White-winged 45, 48, 123
Duck
    Masked 48, 71

**E**

Eagle
    Harpy 105
    Solitary 124
Egret
    Cattle 61
    Great 30, 48, 87, 113
    Snowy 30, 48, 87, 113
Elaenia
    Forest 106
    Gray 106
    Greenish 45, 49, 106
    Lesser 72, 133
    Mountain 60, 64, 69, 75, 130,135
    Yellow-bellied 45, 49, 72, 87, 133
Emerald
    Coppery-headed 74
    Fork-tailed 45, 49, 51
    Garden 35, 39
    White-tailed 124, 129, 132

Euphonia
  Blue-hooded 59, 125, 131, 133
  Fulvous-vented 107
  Olive-backed 36, 56, 63, 81, 85
  Scrub 46, 50
  Spot-crowned 40, 129, 131
  Tawny-capped 56, 59, 63, 66, 119, 121, 125, 133
  Thick-billed 29, 46, 50, 72, 103, 119, 125
  White-vented 36, 40, 107, 121, 125
  Yellow-crowned 29, 33, 36, 40, 81, 85, 103
  Yellow-throated 76, 87
F
Fairy
  Purple-crowned 28, 32, 35, 39, 55, 62, 66, 105, 124
Falcon
  Bat 34, 38, 65, 132
  Laughing 34, 38, 44, 48, 101, 135
Finch
  Large-footed 60, 65, 69, 131
  Peg-billed 69
  Slaty 134, 135
  Sooty-faced 59, 66, 76, 126, 134
  Yellow-green 134, 135
  Yellow-thighed 60, 65, 69, 76, 131
Flatbill
  Eye-ringed 28, 35, 40, 56, 62, 125
  Olivaceous 106, 115, 125
Flower-piercer
  Slaty 64, 69, 75, 134, 135
Flycatcher
  Acadian 102, 125
  Black-capped 60, 64, 69, 130
  Black-tailed 29, 106, 125
  Boat-billed 45, 49, 102
  Bran-colored 133
  Brown-crested 45, 49, 51
  Brownish 106, 125
  Dusky-capped 29, 36, 40, 45, 49, 62, 72, 75, 85, 102, 106, 119, 121, 125, 130, 133
  Fork-tailed 102, 114, 123
  Golden-bellied 58, 66, 69, 130, 133
  Great Crested 32, 45, 49, 51

  N. Scrub 123
  Nutting's 45, 49, 51
  Ochre-bellied 28, 35, 39, 56, 62, 72, 81, 106, 115, 119, 121
  Olive-striped 56, 62, 75, 121, 125
  Panama 102, 114
  Piratic 45, 49, 62, 72, 102, 130
  Royal 29, 106
  Ruddy-tailed 29, 35, 40, 80, 102, 106, 115, 125
  Rusty-margined 102, 109, 114
  Scissor-tailed 46, 49
  Slaty-capped 62, 66, 121, 125, 133
  Social 45, 49, 62, 102
  Streaked 32, 45, 49, 51, 102, 130
  Sulphur-bellied 45, 49, 75
  Sulphur-rumped 35, 40, 106, 121
  Tufted 56, 58, 66, 69, 75, 121, 125, 130, 133
  White-ringed 80, 106
  Yellow-bellied 29, 32, 66, 80, 121
  Yellowish 56, 62, 69, 75, 133
  Yellow-margined 56, 66, 80, 102, 106, 121
  Yellow-olive 28, 32, 45, 49, 51, 56, 62, 72, 102
Foliage-gleaner
  Buff-throated 28, 72, 80, 106
  Lineated 75, 124, 132
  Ruddy 72, 130, 132
  Slaty-winged 106, 120
  Spectacled 62, 130, 132
Forest-Falcon
  Barred 74, 105, 120, 132
  Collared 105
  Slaty-backed 105
Frigatebird
  Magnificent 30, 44
Fruitcrow
  Purple-throated 80, 106, 125
G
Gallinule
  Purple 48, 71, 87, 113
Gnatcatcher
  Tropical 29, 32, 36, 40, 56, 62, 81, 85, 102, 107, 125
  White-lored 46, 49, 51

175

Gnatwren
    Long-billed 29, 32, 36, 40, 81, 85, 102, 107, 121, 125
    Tawny-faced 56, 66, 81, 107, 125
Goldentail
    Blue-throated 28, 32, 35, 39, 45, 134
Goldfinch
    Lesser 59, 103, 114, 131, 135
Grackle
    Nicaraguan 86, 87
Grass-Finch
    Wedge-tailed 118, 119
Grassquit
    Blue-black 29, 72, 85, 114
    Yellow-faced 46, 50, 59, 63, 72, 76, 81, 119, 126, 131
Grebe
    Least 48
    Pied-billed 48
Greenlet
    Golden-fronted 102
    Lesser 29, 32, 36, 40, 46, 50, 51, 56, 62, 66, 75, 81, 85, 102, 107, 119, 125, 130, 133
    Scrub 36, 40, 102
    Tawny-crowned 36, 40, 81, 107, 125
Grosbeak
    Black-faced 56, 81, 122
    Black-thighed 76, 134
    Blue-black 29, 33, 36, 40, 81, 85, 119, 126
    Rose-breasted 46, 50, 63, 66, 85
    Slate-colored 56, 66, 81, 107, 122, 126
Ground-Cuckoo
    Lesser 45, 49, 51
    Rufous-vented 105
Ground-Dove
    Blue 28, 32, 45, 48, 84, 113
    Common 45, 48
    Plain-breasted 123
    Ruddy 61, 113
Ground-Sparrow
    White-eared 76
Guan
    Black 58, 64, 74, 120, 129, 132
    Crested 28, 34, 44, 55, 79, 105, 132

Gull
    Laughing 30, 32, 34, 38, 45
**H**
Hawk
    Barred 120, 124, 131
    Black-chested 58, 64, 65, 74
    Black-collared 87
    Broad-winged 28, 51, 84, 101, 105
    Gray 28, 44, 48, 51
    Harris' 44, 48
    Plumbeous 105
    Red-tailed 64, 69, 132
    Roadside 32, 44, 48, 51, 101
    Savannah 101
    Semiplumbeous 79, 105
    Short-tailed 120, 124, 131
    Tiny 105
    White 34, 38, 55, 105, 124
    White-tailed 44, 48, 119
    Zone-tailed 44, 48
Hawk-Eagle
    Black 61, 79, 105, 124, 132
    Ornate 132
Hermit
    Bronzy 28, 35, 39, 80, 84
    Green 55, 58, 62, 65, 74, 115, 120, 124, 129, 132
    Little 28, 32, 35, 39, 65, 72, 80, 84, 105, 119, 120, 124, 132
    Long-tailed 28, 35, 39, 80, 84, 101, 105, 120, 124
    Rufous-breasted 101, 105
Heron
    Agami 104, 105
    Boat-billed 30, 48, 113
    Capped 113
    Cocoi 113
    Great Blue 30, 48, 87, 113
    Green 48, 71, 87, 113
    Little Blue 30, 87, 113
    Striated 113
    Tricolored 30, 87
Honeycreeper
    Green 29, 56, 63, 72, 103, 115, 119, 121, 125
    Red-legged 29, 33, 36, 40, 46, 50, 85, 103, 121

Shining 33, 36, 40, 56, 81, 107, 119, 121, 125
Hummingbird
    Black-bellied 58
    Blue-chested 80, 84
    Charming 32, 35, 39
    Cinnamon 45, 49, 51
    Fiery-throated 60, 64, 69, 74, 129
    Glow-throated 134, 135
    Magnificent 64, 69, 75, 132, 135
    Mangrove 30, 35, 39
    Ruby-throated 32, 45, 49, 51
    Rufous-tailed 28, 32, 35, 39, 45, 49, 62, 72, 84, 87, 101, 124
    Scaly-breasted 28, 35
    Scintillant 60, 64, 75, 129
    Snowy-bellied 101, 119, 124
    Steely-vented 45, 49, 51
    Stripe-tailed 75, 132
    Violet-bellied 101
    Violet-capped 115, 120
    Violet-headed 35, 39, 55, 66, 115, 124
    Volcano 60, 64, 69, 129

**I**
Ibis
    Glossy 48
    White 30, 48, 87, 113

**J**
Jabiru 46, 48
Jacamar
    Great 105
    Rufous-tailed 28, 35, 39, 80
Jacana
    Northern 84
    Wattled 101, 113
Jacobin
    White-necked 35, 39, 80, 101, 105, 124
Jay
    Azure-hooded 66, 75, 133
    Black-chested 102, 121, 125, 130
    Brown 62, 75
    Silvery-throated 69
Junco
    Volcano 60

**K**
Kestrel
    American 44, 48
Kingbird
    Tropical 45, 62
    Western 46
Kingfisher
    Am. Pygmy 35, 39, 99, 109, 114
    Amazon 87, 114
    Belted 84
    Green 32, 80, 84, 87, 105, 114, 123
    Green-and-Rufous 86, 109, 114
    Ringed 32, 35, 39, 84, 87, 101, 114
Kiskadee
    Great 45, 49, 62, 102
    Lesser 109, 114
Kite
    Double-toothed 44, 48, 61, 105, 120
    Gray-headed 105
    Hook-billed 108
    Pearl 113
    Plumbeous 44, 48
    Snail 48
    Swallow-tailed 64, 65, 69, 71, 74, 119, 120, 124, 135
    White-tailed 44, 48, 101, 113

**L**
Lancebill
    Green-fronted 58, 132
Lapwing
    Southern 113
Leaftosser
    Scaly-throated 35, 39, 106
    Tawny-throated 106, 132
Limpkin 48

**M**
Macaw
    Scarlet 28, 33, 34
Magpie-Jay
    White-throated 46, 49
Manakin
    Blue-crowned 29, 36, 40, 102, 107
    Golden-collared 32, 36, 40, 102, 121, 125
    Lance-tailed 102, 125
    Long-tailed 46, 49, 51, 75
    Orange-collared 29

Red-capped 29, 32, 36, 40, 80, 102, 107, 125
Thrushlike 29, 36, 40, 106, 125, 133
White-collared 80, 85, 87
White-crowned 56, 133
White-ruffed 56, 62, 66, 72, 119, 121, 125, 133
Mango
    Black-throated 114
    Green-breasted 45, 49
Martin
    Gray-breasted 29, 32, 46, 49, 85, 87, 102, 107, 114, 130
Meadowlark
    Eastern 33, 69, 76, 114, 131
Mockingbird
    Tropical 115
Moorhen
    Common 48, 71
Motmot
    Blue-crowned 35, 39, 45, 51, 62, 72, 101, 105, 129
    Broad-billed 55, 62, 80, 101, 105
    Keel-billed 55
    Rufous 62, 80, 101, 105, 120, 124
    Turquoise-browed 45, 49, 51
Mountain-gem
    Variable 58, 69, 75, 120, 124, 129, 132, 135
    White-bellied 58, 124, 132
Mourner
    Rufous 29, 36, 40, 62, 80, 121, 125
    Speckled 106, 121
Muscovy 48, 87
N
Nighthawk
    Lesser 28, 45, 49, 51, 87, 101
    Short-tailed 105
Night-Heron
    Black-crowned 48, 113
    Yellow-crowned 30, 48
Nightingale-Thrush
    Black-billed 60, 64, 69, 130
    Black-headed 56, 75
    Orange-billed 58, 125
    Ruddy-capped 64, 69, 75, 130
    Slaty-backed 66, 72, 75, 121, 125

Nightjar
    Dusky 64, 69, 129
    Rufous 115
Nunbird
    White-fronted 80
O
Oriole
    Black-cowled 85
    Northern 29, 33, 46, 50, 76, 85, 103
    Orchard 85, 103
    Spot-breasted 46
    Streak-backed 46, 50
    Yellow-backed 103, 107
Oropendola
    Chestnut-headed 85, 107, 126
    Crested 103, 107, 126
    Montezuma 63, 81, 85, 87
Osprey 44, 48, 87
Ovenbird 81
Owl
    Barn 45, 49, 114
    Black-and-White 105, 132
    Crested 105
    Mottled 35, 39, 45, 49, 105, 129, 132
    Spectacled 35, 39, 79
P
Parakeet
    Barred 64, 69
    Crimson-fronted 28, 61, 71,84, 132
    Olive-throated 79, 87
    Orange-chinned 28, 32, 34, 38, 45, 49, 71, 105, 115, 123
    Orange-fronted 45, 49, 51
    Sulphur-winged 69, 129, 132
Parrot
    Blue-headed 83, 84, 105, 120, 132
    Brown-hooded 34, 39,58, 79, 124
    Mealy 28, 34, 39, 79, 84, 105
    Red-lored 28, 32, 74, 79, 101, 105
    White-crowned 28, 32, 34, 39, 55, 58, 61, 65, 69, 71, 79, 87
    White-fronted 45, 49, 51
    Yellow-crowned 101, 113, 123
    Yellow-naped 28, 45, 49
Parrotlet
    Blue-fronted 115
    Red-fronted 120

178

Parula
  Tropical 59, 63, 66, 72, 125,133
Pauraque 32, 35, 39, 45, 49, 51, 62, 79, 119, 124, 132
Pelican
  Brown 30, 44
Peppershrike
  Rufous-browed 58, 69, 75, 130, 135
Pewee
  Dark 58, 69, 75, 130, 133
  Ochraceous 69
  Tropical 62, 80, 85, 87, 102
Phoebe
  Black 58, 130, 133
Piculet
  Olivaceous 35, 39
Pigeon
  Band-tailed 58, 60, 64, 69, 74, 129, 132, 135
  Pale-vented 32, 34, 38, 84, 87, 101
  Red-billed 45, 48, 60, 61
  Ruddy 58, 74, 132
  Scaled 32, 71, 101, 115, 129, 135
  Short-billed 28, 34, 38, 55, 61, 79, 84, 105, 120, 124
Piha
  Rufous 29, 36, 40, 80, 106, 133
Pintail
  Northern 48
Plover
  Black-bellied 30, 45, 48
  Wilson's 34, 38
Plumeleteer
  Bronze-tailed 80, 84, 124
  White-vented 101, 105
Potoo
  Gray 105
  Great 105
Puffbird
  Black-breasted 105, 108
  Pied 80, 108
  White-necked 35, 39, 45, 101, 105
  White-whiskered 28, 35, 39, 80, 105
Pygmy-Owl
  Andean 69
  Ferruginous 32, 45, 49, 120
  Least 79

Pygmy-Tyrant
  Black-capped 80, 106
  Pale-eyed 102, 134
  Scale-crested 35, 40, 66, 72, 115, 119, 125, 133
Q
Quail-Dove
  Buff-fronted 69, 74, 129
  Chiriquí 74, 129, 132
  Olive-backed 79, 84, 105
  Purplish-backed 55, 119, 120
  Ruddy 34, 38, 105
Quetzal
  Resplendent 64, 69, 75, 129, 132
R
Redstart
  American 72, 121
S
Sabrewing
  Violet 62, 75, 129, 132
Saltator
  Black-headed 63, 81, 85, 107
  Buff-throated 29, 33, 36, 46, 50, 63, 72, 76, 87, 103, 122, 131
  Grayish 46, 50
  Streaked 72, 103, 114, 122, 134
Sanderling 34, 45
Sandpiper
  Least 30, 34, 38, 87
  Semipalmated 30
  Solitary 87
  Spotted 30, 87, 132
  Western 30
Sapoya
  Broad-billed 106
Screech-Owl
  Bare-shanked 64, 69, 74, 132
  Pacific 45, 49, 51
  Tropical 132
  Vermiculated 79, 105
Scythebill
  Brown-billed 56, 120, 124, 132
Seedeater
  Blue 122
  Variable 33, 59, 72, 81, 85, 114, 122
  White-collared 29, 33, 46, 50, 85
  Yellow-bellied 36, 40, 114

179

Seed-Finch
    Lesser 36, 40, 81, 85, 87, 126
    Nicaraguan 83
Shoveler
    Northern 48
Shrike-Tanager
    White-throated 36, 40, 122, 126
Shrike-Vireo
    Green 36, 40, 56, 81, 107, 121
Sicklebill
    White-tipped 55, 120
Silky-Flycatcher 133
    Black-and-Yellow 69, 75, 130, 135
    Long-tailed 60, 64, 69, 130, 133
Sirystes 106
Siskin
    Yellow-bellied 131
Skimmer
    Black 30
Snowcap 55, 80, 120, 124
Solitaire
    Black-faced 58, 75, 121, 125, 130, 133, 135
Spadebill
    Golden-crowned 29, 35, 40, 56, 72, 80, 106
    Stub-tailed 45, 49
    White-throated 75, 121, 125, 133
Sparrow
    Black-striped 29, 33, 72, 81, 85, 87, 103, 114, 119, 122, 126
    House 33
    Olive 46, 50, 51
    Orange-billed 29, 33, 36, 40, 56, 63, 72, 81, 107, 122, 126
    Rufous-collared 59, 60, 63, 65, 69, 76, 119, 131, 134, 135
Spinetail
    Pale-breasted 71, 72, 114, 129
    Red-faced 58, 66, 75, 129, 132
    Slaty 35, 39, 80, 87, 129
Spoonbill
    Roseate 30, 48, 87
Starthroat
    Long-billed 28, 35, 39, 72
    Plain-capped 45, 49

Stilt
    Black-necked 30
Stork
    Wood 30, 48, 113
Sunbittern 104, 105, 124
Sungrebe 34, 79, 86, 87
Surfbird 34
Swallow
    Barn 46, 49, 87, 102, 114
    Blue-and-White 62, 69, 75, 119, 121, 135
    Mangrove 29, 46, 49, 87
    N. Rough-winged 75, 130
    S. Rough-winged 32, 75, 85, 130, 133
    White-thighed 107
Swift
    Band-rumped 28, 35, 39, 105, 124
    Chestnut-collared 74, 132
    Gray-rumped 55, 79, 84
    Lesser Swallow-tailed 32, 35, 39
    Short-tailed 101
    Vaux's 32, 58, 62, 64, 65, 72, 74, 124, 129, 132
    White-collared 55, 60, 62, 64, 65, 74, 124, 129

**T**

Tanager
    Bay-headed 56, 59, 63, 66, 72, 119, 121, 125, 130, 133
    Black-and-Yellow 56, 66, 115, 119, 122, 126
    Blue-and-Gold 56, 122, 125
    Blue-gray 46, 50, 63, 72
    Crimson-backed 103, 107
    Crimson-collared 59, 63, 66, 81, 85
    Dusky-faced 56, 81, 107, 122, 126
    Emerald 56, 66, 115, 121, 125
    Flame-colored 60, 63, 131, 134, 135
    Flame-rumped 126
    Golden-hooded 29, 33, 36, 40, 59, 63, 72, 81, 85, 130
    Gray-headed 36, 40, 107
    Hepatic 36, 76, 119, 122, 126, 133, 135
    Olive 56, 81, 115, 119, 122, 126
    Palm 63

Plain-colored 81, 103, 115, 121
Rufous-winged 59
Scarlet-rumped 59, 63, 66, 85
Silver-throated 56, 59, 63, 66, 72, 76, 119, 121, 125, 130, 133
Spangle-cheeked 59, 76, 130, 133, 135
Speckled 56, 66, 72, 115, 121, 125
Sulphur-rumped 85
Summer 29, 33, 36, 40, 46, 50, 56, 63, 72, 81, 85, 122, 134
Tawny-crested 56, 81, 85, 119, 122, 126
Western 46, 50
White-lined 85, 122
White-shouldered 36, 40, 56, 63, 85, 103, 107
White-winged 59, 131, 134
Tapaculo
    Silvery-fronted 64, 69, 75, 130, 135
Teal
    Blue-winged 48, 71, 87, 101, 113
Tern
    Black 48
    Elegant 30, 32
    Royal 30, 32, 45, 84
    Sandwich 30, 32
Thorntail
    Green 58, 66, 120, 132
Thrush
    Clay-colored 46, 49, 62, 75, 85, 102, 121, 125, 130, 133
    Mountain 64, 69, 75, 130, 133, 135
    Pale-vented 56, 66, 121, 125, 133
    Sooty 60, 64, 69, 130
    Swainson's 46, 49, 56, 75, 107, 121, 125
    White-throated 72, 75, 119, 121, 125, 133
    Wood 56, 66
Thrush-Tanager
    Rosy 103
Tiger-Heron
    Bare-throated 48, 113
    Fasciated 131
    Rufescent 86

Tinamou
    Great 28, 34, 38, 55, 79, 105, 124
    Highland 69
    Little 34, 38, 61, 79, 84, 101, 120, 124, 131, 135
    Slaty-breasted 79
    Thicket 44, 48, 51
Tityra
    Black-crowned 32, 46, 62, 102
    Masked 46, 49, 62, 75, 85, 102, 130
Tody-Flycatcher
    Black-headed 80, 85
    Common 28, 45, 49, 72, 80, 85, 87, 102, 106, 114
    Slaty-headed 80
Toucan
    Chestnut-mandibled 35, 39, 56, 80, 102, 105, 124
    Keel-billed 45, 49, 62, 80, 84, 102, 105, 115, 119, 124
Toucanet
    Emerald 58, 62, 64, 69, 75, 120, 124, 129, 132
    Yellow-eared 55, 120, 124
Treehunter
    Streak-breasted 69, 75, 130, 132
Treerunner
    Ruddy 64, 69, 75, 130, 132, 135
Trogon
    Baird's 28, 35, 39
    Black-headed 28, 45, 49, 51, 87
    Black-tailed 109
    Black-throated 35, 39, 55, 80, 84, 101, 105
    Collared 58, 64, 66, 69, 129
    Elegant 45, 49, 51
    Lattice-tailed 55, 120, 124
    Orange-bellied 75, 119, 120, 124, 132, 135
    Slaty-tailed 28, 35, 80, 101, 105, 120
    Violaceous 28, 35, 39, 45, 49, 55, 62, 72, 80, 84, 101, 105
    White-tailed 105
Tuftedcheek
    Buffy 58, 64, 75, 130, 132
Turnstone
    Ruddy 30

Tyrannulet
    Brown-capped 81, 102
    Mistletoe 3
    Mouse-colored 102, 114
    N. Beardless 45, 49
    Paltry 35, 39, 58, 62, 66, 72, 75, 81, 102, 121, 125, 130, 133
    Rufous-browed 56, 66, 125
    S. Beardless 28, 32, 35, 39, 102, 114, 125
    Torrent 66, 130, 133
    White-fronted 69, 133
    Yellow 35, 40, 102
    Yellow-bellied 35, 39, 87
    Yellow-crowned 102
Tyrant
    Long-tailed 80, 85, 106, 125

## U
Umbrellabird
    Bare-necked 133

## V
Violetear
    Brown 65, 120, 124
    Green 60, 62, 64, 69, 75, 132, 135
Vireo
    Brown-capped 58, 62, 64, 69, 75, 130, 133, 135
    Mangrove 30, 31
    Philadelphia 32, 133
    Yellow-green 29, 46, 51, 62, 75, 102
    Yellow-throated 29, 51, 62, 66, 85, 102, 115, 133
    Yellow-winged 64, 69, 130
Vulture
    Black 61
    King 34, 38, 44, 48, 105
    Lesser Yellow-headed 48, 87, 113
    Turkey 61

## W
Warbler
    Bay-breasted 85, 121
    Black-and-White 59, 63, 66, 75, 121
    Blackburnian 59, 63, 66, 75, 81, 121, 125, 130, 133
    Black-cheeked 65, 69, 130, 135
    Black-throated Green 59, 65, 66, 69, 75, 130, 133, 135
    Buff-rumped 36, 40, 56, 63, 81, 121, 125
    Canada 36, 40, 75, 133
    Chestnut-sided 29, 32, 56, 59, 63, 66, 81, 85, 121, 130, 133
    Flame-throated 60, 64, 69, 130
    Golden-crowned 63, 66, 76, 125, 130
    Golden-winged 58, 63, 66, 72, 75, 121, 125, 130, 133
    Kentucky 33, 36, 40, 63, 66, 72, 81, 85, 107
    MacGillivray's 133
    Mangrove 31, 99, 109, 123
    Mourning 36, 40, 59, 63, 66, 121, 125, 130
    Prothonotary 32, 36, 40, 85, 103, 123
    Rufous-capped 46, 50, 51, 63, 72, 130, 135
    Tennessee 29, 32, 36, 40, 46, 50, 51, 58, 63, 85, 121, 130, 133
    Three-striped 59, 66, 76, 125, 133
    Townsend's 75
    Wilson's 59, 60, 63, 65, 66, 69, 72, 130, 133, 135
    Worm-eating 33
    Yellow 46, 50, 87
Waterthrush
    Louisiana 121, 133
    Northern 46, 50, 81, 85, 87, 107, 123
Water-Tyrant
    Pied 114
Whimbrel 30, 45
Whistling-Duck
    Black-bellied 48, 87
    Fulvous 48
Whitestart
    Collared 59, 65, 69, 75, 130, 135
    Slate-throated 59, 60, 63, 66, 69, 76, 121, 125, 130, 133, 135
Wigeon
    American 48
Willet 30, 45, 48
Woodcreeper
    Barred 28, 80, 84, 106
    Black-striped 35, 39, 80, 106
    Buff-throated 28, 35, 39, 80, 84, 106

Ivory-billed 45, 49
Long-tailed 35, 39, 106
Olivaceous 62, 75, 115, 132
Plain-brown 80, 84, 106, 120, 124
Ruddy 45
Spot-crowned 64, 69, 130
Spotted 56, 66, 75, 120, 124, 132
Straight-billed 99, 109
Streak-headed 28, 32, 35, 39, 45, 49, 62, 66, 84, 109, 132, 135
Strong-billed 132
Tawny-winged 28, 32, 35, 39
Wedge-billed 28, 35, 39, 56, 80
Woodhaunter
    Striped 35, 124
Woodnymph
    Violet-crowned 28, 35, 39, 55, 62, 66, 80, 101, 105
Wood-Partridge
    Buffy-crowned 60
Woodpecker
    Acorn 60, 64, 69, 129, 132, 135
    Black-cheeked 56, 62, 80, 84, 102, 106, 115
    Chestnut-colored 80
    Cinnamon 56, 80, 106, 120, 124
    Crimson-bellied 106
    Golden-naped 28, 32, 35, 39
    Golden-olive 62, 75, 120, 124, 132
    Hairy 64, 69, 75, 129
    Hoffman's 45, 49, 51
    Lineated 28, 45, 49, 62, 84, 102, 106
    Pale-billed 28, 35, 45, 49, 51, 56, 80
    Red-crowned 32, 35, 39, 72, 102, 114
    Rufous-winged 39, 56, 62, 66, 80, 124
    Smoky-brown 56, 66, 75, 124, 132
    Spot-breasted 99
Wood-Quail
    Black-breasted 74, 132
    Marbled 34, 38, 105
    Rufous-fronted 55
    Spotted 64, 129
Wood-Rail
    Gray-necked 32, 34, 38, 44, 48, 84, 87, 105

Woodstar
    Magenta-throated 75, 132
Wood-Wren
    Gray-breasted 58, 62, 64, 69, 75, 119, 121, 125, 130, 133, 135
    White-breasted 36, 40, 56, 81, 102, 107, 119, 121, 125
Wren
    Band-backed 56, 62, 81, 85, 125
    Banded 46, 49, 51
    Bay 56, 62, 66, 81, 107, 121, 125
    Black-bellied 29, 32, 36, 40, 107
    Black-throated 85
    Buff-breasted 102
    House 29, 46, 49, 62, 69, 75, 114
    N. Nightingale 66
    Ochraceous 58, 64, 69, 75, 130, 133, 135
    Plain 29, 32, 36, 40, 46, 49, 62, 72, 85, 130, 133, 135
    Riverside 29, 32, 36, 40
    Rufous-and-White 46, 49, 75, 125
    Rufous-breasted 29, 72, 121, 125, 130
    Rufous-naped 46, 49, 51
    S. Nightingale 36, 40, 72, 107, 121, 125, 133
    Song 81, 102, 107, 119
    Stripe-breasted 56, 62, 81, 121, 125
    Timberline 69, 130
    White-headed 108
Wrenthrush 65, 69, 76, 133
X
Xenops
    Plain 28, 35, 39, 56, 62, 102, 106, 119, 120, 124
Y
Yellowlegs
    Lesser 87
Yellowthroat
    Chiriquí 71
    Gray-crowned 36, 40, 46, 50, 81
    Masked 71, 72, 130
    Olive-crowned 81, 85

# Also From Cinclus Books

## *Finding Birds in Britain: A Site Guide*
By Graham Speight  1994
Written with the visiting birder in mind, this book gives complete coverage along with maps and species lists for 43 sites in England, Scotland, and Wales.
ISBN 0-937765-4-1  140 p. paper  $14.50

## *Site Guides: La Ruta Maya*
*A Guide to the Best Birding Locations in*
*The Yucatan, Belize, Guatemala, Honduras, and El Salvador*
By Dennis Rogers  1994
Birding guide to the former empire of the Maya, with information on visiting each country involved and details and species lists for each specific site.
ISBN 0-967765-3-3  54 p. paper  $8.95

## *Site Guides: Venezuela*
*A Guides to the Best Birding Locations*
By Dennis Rogers    1993
Coverage of 20 prime birding locations in all parts of the country, with possible species for each. A complete checklist of all 1350 species in Venezuela is included, along with an index of all species mentioned in the text.
ISBN 0-937765-0-9  48 p.  8½X11 spiral bound $14.50

Site Guides: Costa Rica & Panama ___ @ $19.95  ___
Finding Birds in Britain: A Site Guide ___ @ $14.50  ___
Site Guides: La Ruta Maya ___ @ $8.95  ___
Site Guides: Venezuala ___ @ $14.50  ___
Please add $2 S&H for first book, $1 each additional  ___

Order from:     Cinclus S.A.          Total  ___
                Box 80414
                Portland, OR, 97219